# Manufacturing Systems Redesign

## Creating the Integrated Manufacturing Environment

**David O'Sullivan**

*University College-Galway, Ireland*

P T R Prentice Hall
Englewood Cliffs, New Jersey 07632

**Library of Congress Cataloging-in-Publication Data**
O'Sullivan, David.
    Manufacturing systems redesign : creating the integrated
manufacturing environment / David O'Sullivan.
        p. cm.
    Includes bibliographical references and index.
    ISBN 0-13-072786-5
    1. Computer integrated manufacturing systems--Planning.
2. Production planning. 3. Flexible manufacturing systems-
-Planning. I. Title.
TS155.63.O88 1994
658.5--dc20
                                        94-8916
                                        CIP

Editorial production and interior design: **Ann Sullivan**
Cover designer: **Tom Nery**
Cover photo: **The Image Bank (211050)/Garry Gay**
Manufacturing buyer: **Alexis R. Heydt**
Acquisitions editor: **Bernard Goodwin**
Editorial assistant: **Diane Spina**

©1994 by P T R Prentice Hall
Prentice-Hall Inc.
A Division of Simon & Schuster
Englewood Cliffs, NJ 07632

The publisher offers discounts on this book when ordered in bulk quantities.
For more information, contact:

    Corporate Sales Department
    P T R Prentice Hall
    113 Sylvan Avenue
    Englewood Cliffs, NJ 07632
    Phone: 201-592-2863
    FAX: 201-592-2249

Printed in the United States of America
10 9 8 7 6 5 4 3 2 1

**ISBN   0-13-072786-5**

Prentice-Hall International (UK) Limited, *London*
Prentice-Hall of Australia Pty. Limited, *Sydney*
Prentice-Hall of Canada Inc., *Toronto*
Prentice-Hall Hispanoamericana, S.A., *Mexico*
Prentice-Hall of India Private Limited, New *Delhi*
Prentice-Hall of Japan, Inc., *Tokyo*
Simon & Schuster Asia Pte. Ltd., *Singapore*
Editora Prentice-Hall do Brasil, Ltda., *Rio de Janeiro*

# Dedication

*To Bernadette and Eóin
and happy years together*

# Contents

# Preface

## INTRODUCTION

Manufacturers are faced with increasing challenges as a result of increased market competition and changing customer demands. Increased market competition is driving manufacturing systems to become more productive and efficient. Increased customer demand for greater product variety is requiring manufacturing systems to become more flexible. Many manufacturing companies are finding that they need to change their manufacturing systems more frequently than ever before to keep pace with these changes. They are also finding that they need to continuously improve the manufacturing process to remain competitive.

The people responsible for these improvements—the manufacturing system design group—are also finding their environment changing and increasing in complexity. Rapid changes in technology, which now involve the use of more sophisticated computer based hardware and software, are demanding new levels of competence from systems designers and users. Integration of systems into their environment is now seen as essential to capture the outer limits of productivity. Finally, changes in the social environment within manufacturing are forcing significant changes to the manufacturing organization and the roles of people. To help designers cope, advanced tools and methodologies are required to augment their design skills and provide a systematic approach to the design process.

This book is about the development and use of a technique called the Integrated Manufacturing Systems Design Procedure (IMP), which is a procedure for augmenting the activities of the manufacturing systems design group in developing and implementing manufacturing projects. The procedure consists of two parts—a theoretical part and an implementable part. The theoretical part of IMP is developed on a number of ideas about the way that manufacturing projects can be implemented, and in particular on a set of seven principles that are created to mold the way in which designers think about manufacturing systems. The implementable part of IMP is designed to facilitate these new ideas and principles and to provide an unambiguous road map for the design group in the design of manufacturing projects. The IMP procedure augments the skills of the designer and facilitates the integration of designed systems into their technical and social environments.

## Intended Audience

This book is primarily written for members of manufacturing project groups who are responsible for the design of manufacturing systems. The intention is that the book can be used as a road map for the mechanisms used in the organization, planning, development, and implementation of these complex systems. Manufacturing managers will find the book useful for understanding the activities of manufacturing systems design and in particular the interaction between manufacturing systems design and manufacturing strategy. Manufacturing system designers will find the book useful for understanding the underlying theory of systems design, as well as the mechanisms and resources of systems design used at various stages of a project life cycle. The book will also be useful for applications-oriented postgraduate courses in manufacturing systems design. The approach adopted in the book is systematic, and students of manufacturing will find the models presented helpful for their understanding of the manufacturing environment. Surveys have shown that relatively few manufacturing systems are currently being implemented efficiently and cost effectively. It is hoped that this book will in some way enable members of project groups to address this problem and enable them to develop a continuous change process for manufacturing.

## Book Outline

Manufacturing systems incorporate two distinct areas of study—manufacturing operations and manufacturing systems design. Many books have been written about manufacturing operations, and the ideal manufacturing environment or factory of the future. These books promote ideas about new types of operation paradigms for manufacturing in areas such as production management systems, concurrent engineering, and shop-floor control. In contrast, my book focuses on manufacturing systems design. In it, I discuss the design approach to manufacturing systems that may or may not involve many of the modern operations paradigms. As such it deals mainly with issues such as manufacturing strategy, systems design programs, stages in project implementation, and tools for facilitating systems design.

Traditional descriptions of the manufacturing systems design process have been limited to issues such as project management, the use of Gantt charts, and cost-benefit analysis. In recent years however, design issues have exploded in scope and depth. It is the purpose of this book to explore many of these issues and to put forward a design approach that can help designers to cope with them. The book is broadly structured into three areas. The first area, covered mainly in Chapter 1 examines the context for manufacturing systems design. The second area, dealt with in Chapters 2 and 3, describes the various social and technical planning tools available for facilitating designers in the manufacturing systems design process. These chapters are essentially reviews of existing technical and social planning tools. The reader may decide to skip these chapters initially and return to them when using the IMP road map. The third area is an approach to manufacturing systems design. This approach, called IMP, begins with an understanding of the theory of manufacturing systems design and concludes with the presentation of a design road map for guiding systems designers through the various stages and interactions of systems design. This third area is dealt with in Chapters 4 and 5. Chapter 6 consists of a number of case studies partially implemented using the so-called IMP approach. These case studies are also used to illustrate further many of the procedural aspects of IMP. Each chapter will now be described in more detail.

Chapter 1 sets the context for the activities of manufacturing systems design. Many of the pressures facing manufacturing systems designers are outlined. A description of manufacturing systems is then explored, first in the context of open systems theory and seconds from two additional viewpoints. The first viewpoint covers the functions involved in a typical manufacturing environment. The second viewpoint describes how these functions are linked through various strategies and goals. One of these goals, *integration* is then explored in more detail. This chapter concludes with a description of the modeling technique IDEFo which is used throughout the book for representing ideas and information.

In Chapters 2 and 3 various technical and social planning tools, techniques, and methodologies are reviewed for their contribution to manufacturing systems design. Chapter 2, *Technical Planning Systems*, explores areas such as standards, systems architectures, systems design tools, and methodologies. These tools are used increasingly for the design of manufacturing subsystems such as information flow, data structure, and project implementation. Chapter 3, *Social and Systems Planning*, looks at the less conclusive areas of systems design, such as organizational development, group dynamics, and a number of tools and methodologies for designing the social subsystem in manufacturing. Chapter 3 concludes with a framework illustrating where many of the technical and social planning tools discussed are applied in the manufacturing systems design process.

In Chapter 4, an approach to understanding a theory of manufacturing systems design is made. In manufacturing, theory is often seen as an unnecessary intrusion into the understanding of systems design, where often intuition and experience are the essential requirements. However, in the current complexities of manufacturing, where change is occurring rapidly, experience is becoming harder to find. In addition, it is clear to many good systems designers that there are attributes, factors, and approaches that are common in all design projects and that, if formulated and mapped, can be passed on to other less experienced designers. Chapter 4 investigates these factors and attempts to put together a theoretical framework for manufacturing systems design. The results of this investigation lead directly to the creation of the integrated manufacturing systems design road map or IMP.

Chapter 5 describes the integrated manufacturing systems design road map. The road map is a structured view of many of the issues and ideas discussed in previous chapters and is laid out so that designers can map out their own approach to systems design. It consists of text and an IDEFo model. The use of the model allows detailed information to be structured and unambiguously represented. The road map begins by bringing the systems designer through the various activities of systems design from context development through to individual project stages. In each activity key resources are indicated as well as key input and output flows. Each activity also illustrates the key design constraints and any feedback and control information to other activities.

Chapter 6 presents three case studies partially implemented using many of the ideas and techniques discussed in earlier chapters and approached using the road map described in Chapter 5. The first case study involves the development of a human centered flexible assembly system; the second examines a sheet metal flexible manufacturing system; the third case study describes the development of a software package for manufacturing systems design. In conclusion, Chapter 6 outlines some of the advantages and benefits of the IMP approach. It also outlines new changes facing manufacturing in the future from an operational perspective and discusses how these changes will impact on the systems design process. A diagrammatic representation of each of the chapters and their association is given in Figure 1.

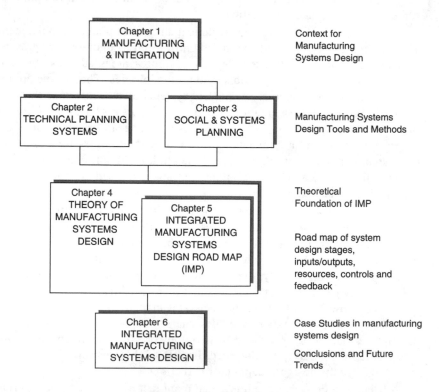

**Figure 1**   Structure of the book

# Acknowledgements

I would like to thank all of my friends and colleagues at University College Galway, Thermo King Europe Ltd., and Laboratorie GRAI of Bordeaux for their contributions to many of the issues discussed in this book. I would like to thank in particular Jim Browne, John Roche, and John Lane for their experience, support, and companionship. Thanks also to three talented young researchers at UCG, William Egenton, Thomas Craven, and Sean O'Connell. I would also like to thank my immediate family for their support, my son Eóin, the O'Sullivan's at 46 (my parents Patrick and Joan), James and Unice, Gemma and Roberta, and Jude and Paul Hulgraine, as well as my extended family throughout Ireland, the United States, Canada, the United Kingdom, and Sneem. My most sincere thanks go to my wife and partner Bernadette for her love and support.

# Manufacturing and Integration

## 1.1 INTRODUCTION

One of the most important roles played by any major economy is its ability to manufacture products that satisfy customer demands globally. This ability depends on a number of factors. Within the manufacturing enterprise these factors can loosely be said to depend on abilities within the functional areas of Product Design, Manufacturing, Finance, and Marketing. In the last two decades the factors that have been important in manufacturing have been quality, lead time, and costs. The design of manufacturing systems that are able to meet the challenges provided by such factors requires systems design groups which themselves are able to reach increasingly advanced states of performance. Traditionally two groups within manufacturing have been responsible—Management Information Systems (MIS) and Manufacturing Systems Design (MSD). This book is concerned with the activities of the latter group. This chapter aims to set the context for the manufacturing systems design function in terms of the environment where MSD is carried out and the issues and goals that surround it.

We will begin with an overview of what we understand to be systems. This overview and the concepts that underlie systems theory are critical to our understanding of manufacturing systems design. Later we will look at two views of manufacturing systems that are created to help set the context for the activities of systems design. The first view places the activities of systems design into context within the overall activity of manufacturing. The second view looks at how the policy of change for systems ultimately comes about. In particular this view discusses how strategy can impact on successful systems design and how this strategy is translated through a number of critical success factors. The next section of the chapter focuses on one of these critical success factors—integration. Integration is now seen as an important way of achieving the upper limits of efficiency within manufacturing. We will conclude with the introduction of the systems modeling technique IDEFo, which is used throughout the book for representing various ideas and concepts.

## 1.2    MANUFACTURING ENVIRONMENT

There are increasing pressures on manufacturing companies to offer quality and cost-effective products to an increasingly demanding customer. Three categories of pressure facing manufacturing companies can be identified.[1] These are (1) increased competition, (2) changes in the internal social organization of manufacturing, and (3) increasing sophistication of manufacturing technology, particularly computer-based technology. Skinner argues that in order to remain competitive, a company must continually evaluate and implement change in its manufacturing environment. A number of people are clearly responsible for carrying out change. These include personnel at all levels in the organization, from managing directors right down to the people who make change physically happen on the shop floor. The people who are primarily responsible for implementing change are the manufacturing systems design group. Their own internal environment within manufacturing is also going through major technical and social changes.

A survey carried out by the Society of Manufacturing Engineers in the United States identified four trends that will create unprecedented change for the manufacturing systems designer by the year 2000. The first of these trends is identified in the survey as "an environment exploding in scope."[2] By the year 2000, products will be more sophisticated, will have far greater variety than at present, and will be produced by systems far more complex than today's. The second trend identified is that the roles filled by systems designers will change to cope with the increasing complexity of systems. No longer, it is argued, will it be possible for a so-called industry *champion* to ensure project success. Three new distinct roles will have to be created and integrated into the organization structure. These roles were identified in the survey as that of Manufacturing Strategist, Manufacturing Specialist, and Manufacturing Integrator. A third trend identified in the survey is the need for and the use of more advanced tools for systems design. Expert systems, communications devices, and design methodologies will all form part of the toolbox of the system designer of the future. To cope with the use and application of these advanced tools, the system designer will need to be educated in the use and application of these tools. Thus, he or she will rely on the support of various educational organizations such as universities. Finally, a fourth trend identified in the U.S. survey points out that by the year 2000, manufacturing will have a changed work emphasis. Systems will be recognized as more human centered. Systems design will need more focus on teams rather than on individuals, and the reliance on outside services such as facilitators and consultants will be greater than ever. Figure 1-1 shows the four trends in pictorial format.

With each of these pressures and expected changes within the manufacturing environment, it is perhaps little wonder that project implementation does not currently run as smoothly as industry would like. Over the last 20 years, manufacturing systems design groups have translated substantial investments into new manufacturing systems. Some of these systems have gained much publicity for their contribution to productivity and flexibility. But while some have made headlines in various publications, many more have failed to meet expectations regarding cost, start-up dates, and subsequent performance. One source indicates project failure rates of between 50–75%.[3] Reasons for these failures are numerous, but some occur more frequently than others. The more frequent reasons for projects not living up to expectations have been identified and include such issues as the lack of commitment and support of top management, the excessive pace of adoption, the lack of education and training of personnel, the influence of the amount of work subcontracted, and the relative infrequency and high cost of manufacturing systems projects.[4, 5]

At a more detailed level in the design of manufacturing systems, projects such as the design and implementation of flexible manufacturing systems involve the analysis and design of complex information systems. A number of common problems associated with such projects can be highlighted.[6] Included among these are that often (1) information systems do not match user requirements, (2) have maintenance costs up to four times developments costs, and (3) have over 60% of their errors occur during requirements analysis rather than development and implementation. Connor concludes that the causes of these problems are lack of proper attention to project preplanning, lack of a structured approach to systems design, and poor emphasis on the analysis of the social subsystem. These same causes have been identified by a number of research organizations and have been used to justify the use of so called CASE (Computer Aided Software Engineering) tools by systems design groups.

Faced with these pressures, potential problems, and a changing work environment, companies and particularly systems designers need to reappraise the way manufacturing systems are designed. There is clearly something wrong with the way many projects are currently being implemented. The reasons outlined above give some clues as to how this situation may be improved. Current literature would appear to distinguish between two areas for dealing with these and other problems in systems design. The first of these is the need for the creation of new systems design tools. The second is the need for more focus on the human subsystem in manufacturing. For example, a number of commentators have indicated that manufacturing is currently resourced by a number of outmoded tools for both technical systems design and investment analysis. There is a clear need for more use of modeling tools such as simulation, computer process control, and statistical quality control.[1] In addition to these, there is a need for focus on the use and creation of more *systems design* oriented tools such as better modeling methodologies.[7, 8] This latter category of tool is often referred to as *structured analysis* and represents an area where various types of modeling methodologies have been developed for complex systems design and model representation.

Regarding the reemphasis on the human aspects of systems, an equally large number of commentators see this as a key area for success and one that requires a new approach. On the one hand, there is a need to organize, motivate, and direct people involved in the design of manufacturing systems.[9] On the other hand, there is a need for better sociotechnical-systems design tools for the design of the technical and social subsystems and their integration.[10] Coupled with the need for better social organization and sociotechnical design approaches is the need for more focus on the concept of the learning organization for allowing companies to learn and develop through experience.[5] What is clear from these issues is that there are a large number of perspectives on manufacturing. Each perspective yields its own issues, ideas, and solutions. Finding a common denominator for all perspectives is perhaps the first step in creating an understanding for manufacturing systems design.

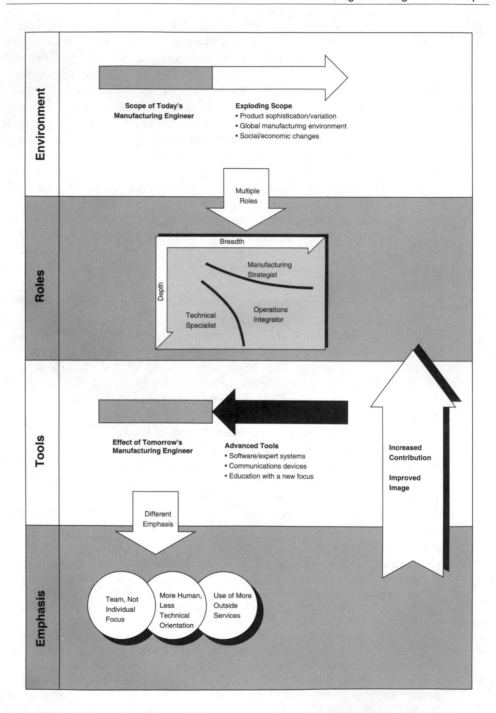

**FIGURE 1-1.** Pictorial results of SME survey (Source: Society of Manufacturing Engineers, 1988, p5.)

## 1.3    OPEN SYSTEMS

In order to study manufacturing systems, their problems, and possible solutions, it is necessary first to have a way to *think about* or conceptualize them. One approach to conceptualization is through the use of simple block diagrams or process flow charts. Although useful in some respects, these types of techniques are highly simplified and only applicable in relatively small problem domains. The approach adopted in this book is to use the techniques and philosophy of systems theory. Systems theory lays the foundation for the analysis of open systems and the way we *think* about systems such as manufacturing.

### 1.3.1 Systems Theory

Systems theory provides the analytical framework for comprehending dynamic, integrated operating systems. The basic concept of systems theory is that all living and many nonliving things are systems that are made up of subsystems and that may be affected by their environments. A popular definition of the term "system" is *an identifiable, complex dynamic entity composed of discernibly different parts or subsystems that are interrelated to and interdependent on each other and the whole entity with an overall capability to maintain stability and to adapt behavior in response to external influences.*

By this definition, clearly all living and many nonliving things are systems, whether they be physical, biological, or social. Close examination of each term gives further insights into the nature of systems, and we can see that manufacturing systems fit easily into this definition. We understand a manufacturing system to be identifiable—perhaps as a building with machinery and people. It is a complex dynamic entity; a manufacturing system is typically made up of a number of parts each of which interact with each other. The interaction occurs in a number of ways. One type of interaction is physical, for example where one type of equipment interacts with another (e.g., conveyor and robot). Another type of interaction is an interaction with respect to time. Various types of equipment can be made to interact with each other for the production of one component through a time schedule (e.g., machine loading). A manufacturing system has interrelated, interdependent discernible different parts—subsystems that interact with each other. For example, the machining subsystem interacts with the assembly subsystem through interactions such as schedules and material flow. A manufacturing system has an overall capability to maintain stability. Under normal operating conditions, a manufacturing system will regulate its various internal subsystems to produce required results. This may mean relocating resources or carrying out maintenance procedures. Finally, a manufacturing system has the ability to adapt its behavior in response to external influences. In many manufacturing systems external changes such as increasing or decreasing customer demand can be detected by the system in advance, and this can then be used to change the behavior of the system (e.g., increasing or decreasing output).

## 1.3.2 General Systems Theory

The concepts underlying systems theory, although traceable back to the sixth century B.C., have only in recent decades been considered a science in their own right. The term used for this science as applied to open systems is General Systems Theory (GST). The use and application of GST and its concept of open systems play a key part in the development of our understanding of manufacturing systems. GST evolved during the 1940s and 1950s in response to biologists, scientists, and economists who saw the need for a newer, more advanced logic capable of embracing studies of both living and nonliving things. GST serves as a framework under which we use generally accepted guidelines to study systems. It is a general theory in that it does not attempt to be either so specific that it sacrifices generality nor so general that it sacrifices content. This aspect of GST is explained by Boulding as:

> "(General Systems Theory) does not seek, of course, to establish a single, self contained *general theory* of practically everything which will replace the theories of particular disciplines. Such a theory would be almost without content, for we always pay for generality by sacrificing content, and we can say about practically everything is almost nothing. Somewhere however between the specific that has no meaning and the general that has no content there must be, for each purpose and at each level of abstraction, an optimum degree of generality. It is the contention of the General Systems Theorists that this optimum degree of generality is not always reached by the particular sciences."[11]

Boulding was one of the founders of the GST society in 1956 and is one of its leading pioneers to date.

The original intention of GST was to present a way in which organic biology could be studied. However, the concepts and approach have become universal across a wide range of disciplines. For example, Boulding was an economist who used GST in economic research. Among his fellow pioneers of the technique were biomathematician Anatol Rapport, biologist and *father* of GST Ludvig von Bertalanffy, and physiologist Ralf Gerard. GST thus provides an analytical framework for the study of any system and encourages the study of relationships within systems more clearly.

### GST Traits

The framework provided by GST is documented through a number of traits pertaining to systems. These traits help us to understand the ways open systems behave. They ultimately help us to analyze and design manufacturing systems. GST is considered a young science in its current state with little unquestionable doctrine. However, the following traits have been attributed to GST beyond the concepts contained in the earlier definition of a *system*. These traits can be said to form a benchmark for the use of GST which system designers may continuously fall back on. Nine such traits can be identified. These state that open systems (1) are goal seeking, (2) are holistic, (3) are hierarchical, (4) have inputs and outputs, (5) transform inputs into outputs, (6) consume or generate energy,(7) are subject to the effects of entropy, (8) have equifinality, and (9) have feedback. Let us make a closer examination of each trait in the context of manufacturing systems design.

*Goal Seeking.* An open system is goal seeking. For example, a manufacturing system always works toward a goal or set of goals. These goals may be defined by the manufacturing or business strategy (outer directed) or they may be defined by a need to perpetuate the life of the system through maintenance (inner directed). Outer directed goals are those goals that contribute to the development of systems in the environment of the system in question. In the case of an entire manufacturing enterprise its goals and strategies would be directed *inward* to perpetuate and change itself and *outward* to satisfy its customers. This concept is captured to some extent by Porter when he writes *"The essence of formulating competitive strategy is relating a company to its environment."*[12]

*Holism.* An open system is holistic, that is, it is an inseparable entity. A manufacturing system for example is comprised of many subsystems. To analyze any of these subsystems it is necessary to examine to some extent the whole. Dissecting the system and analyzing each subsystem individually causes loss of detail essential to the understanding of the system as a whole. Friedrich Hegel, the philosopher (1770–1831) once wrote:

1. The whole is more than the sum of the parts.
2. The whole determines the nature of the parts.
3. The parts cannot be understood if considered in isolation from the whole.
4. The parts are dynamically interrelated and interdependent.

His work is often cited as defining the origin of GST.

*Hierarchy.* An open system has a hierarchy, in which subsystems are nested in a ranking from those of major importance to goal attainment. right down to those subsystems that have minimal effect on goal attainment. In a manufacturing system the assembly line is often regarded as the primary goal-attaining subsystem. An area such as building maintenance on the other hand has perhaps less of an impact on goal attainment. All other subsystems vary accordingly. A phrase often used by a systems design manager is that the main production system must never take second place. In other words, if a problem occurs in the primary goal attainment subsystems, then activities in lesser important subsystems must be sacrificed or put on hold.

*Inputs and Outputs.* An open system depends on inputs and produces outputs. In the case of a manufacturing system, various resources are used as inputs (e.g., raw materials, orders, consumables, people, electricity, money, and oil). Typical outputs are products, scrap, waste, people, and money. It is often important in a design project to identify as many inputs and outputs to a system as possible. Once identified they can be ranked hierarchically according to their importance to the goals of the project. Many inputs and outputs will then be ignored by the project team in preference to the detailed analysis of key inputs and outputs.

*Transformation.* An open system transforms inputs into outputs. In a simplified illustration of manufacturing, the inputs of raw materials and orders are transformed into the output of products. The form of the output is also different from the form of the input.

Within a machining department, a CNC machine will consume three main inputs (NC Code, Raw Materials, and Schedule) and produce three main outputs (Product, Scrap, and Performance Information). Within the broader context of manufacturing the main input to an enterprise can be said to be customer requirements in the form of actual orders and sales forecasts. This information is then translated through marketing, design, and production departments into products that satisfy customer requirements or marketing targets.

*Energy.*    An open system consumes and/or generates energy. In manufacturing this trait is clearly evident. But the term energy is not limited to physical energy such as electricity or manpower. The term energy also includes mental energy such as assertiveness, influence, and spirit. The understanding and utilization of mental energy by a company is clearly a complex issue satisfied mainly through learning and proper management practices. The difference between success and failure in a manufacturing system is often attributable to the proper utilization of mental energy among designers, managers, and operators.

*Entropy.*    An open system is subject to the effects of entropy. This is perhaps the closest link between the loose concepts of GST and what some would regard as the absolute truths of physics. The term entropy is widely used in physics to define the natural tendency of thermodynamic systems to consume themselves unless continuously fed with energy. In manufacturing systems the term is used for the tendency to degenerate unless mental as well as physical energy is continuously fed into the system. The concept of entropy is closely associated with the concept of organizational learning. Organizational learning is defined as "*a system which facilitates the learning of all its members and continuously transforms itself.*"[13] In GST the way in which systems avoid the effects of entropy is *to replace worn equipment and obsolete thinking.*[14] Organizational Learning is a very important trait of manufacturing systems and is discussed in more detail in Chapter 3.

*Equifinality.*    Open systems have equifinality in that they can reach their goals in a number of ways. They have the ability to select and change inputs and to change the ways in which these inputs are processed. This flexibility also allows open systems to avoid the effects of entropy. In manufacturing, flexibility is often designed into a system through people. Line supervisors for example can make decisions on perhaps an hourly basis on which jobs to produce first. Also, based on new information management can decide to change its product mix. The concept of flexibility in manufacturing has evolved continuously over the last 60 years to incorporate a number of different types of flexibility. The following classification for flexibility is popular: machine flexibility incorporating product flexibility, process flexibility, and operation flexibility; routing flexibility incorporating volume flexibility and expansion flexibility; and finally production flexibility which incorporates aspects of all.[15]

*Feedback.*    A final trait that is attributable to open systems under GST is their ability to generate feedback and to use this feedback to change the way they behave. In effect, open systems are servo systems which monitor certain parameters and feed the information back to subsystems, which in turn may alter the behavior of the system. (Note: The terms *open* and *open-loop* are clearly very different. Open-loop systems cannot have feedback, whereas open systems as defined here do). Feedback in manufacturing allows for manufacturing control of product quality, production progress, and maintenance. Increasingly in the design of information systems in manufacturing, feedback of information to management is becoming a larger part of the design specification. This feedback can be status or historical information. Status information is real-time information usually illustrated through the use of icons (which typically show busy and idle machines) or bar charts (showing the status of a schedule). Historical information is typically a report showing system performance. Various types of reports can be produced from detailed analysis (utilization times, store turnover, and quality ratings) to top level management reports showing a summary of various departmental performance indices over a particular time period.

Each of the traits identified through general systems theory help us to understand key attributes of the manufacturing system, albeit at a conceptual level. Understanding these traits is surely one of the first steps to understanding manufacturing systems. One difficulty of course which arises in the minds of a typical design engineer is that they are now suddenly being asked to understand concepts which explore relationships rather than facts, to understand the general rather then the specific. For many systems designers be they experts in software, computer architectures, or processing machines this can be a difficult adjustment from their own reality and experience. Wolf outlines the problem well:

> "A major problem inherent in (the use of systems theory to explain organizational behavior) is that one must accept a concept of relativity that is tremendously complex and, when followed through logically, leaves one with little firm ground upon which to stand. Absolutes vanish, and events and happenings have to be explained in relation to other aspects of the situation."[16]

However, such an approach is important in order to provide a truly open systems approach to design. As we shall see later, the initial vagueness is transitional and once the designer has a general understanding of a systems behavior in terms of many of these traits he or she can narrow the analysis to the details of the subsystem while maintaining his or her overall systems perspective.

## System Classification

An important way to conceptualize the way in which GST can be applied to systems is to classify systems according to various levels of abstraction. This type of classification serves to illustrate the openness and complexity of systems. In the discussion above a system was defined generally as a group of interrelated, interdependent, and interacting subsystems. This general definition was then extended to include various traits for systems according to GST. Two additional classifications of systems are available for providing additional information. Checkland[17] provides a helpful classification when he distinguishes between:

1. Natural Systems (e.g., ecological systems, human body)
2. Physically Designed Systems (e.g., bridges, machines)
3. Abstract Design Systems (e.g., languages, mathematics)

**4.** Human Activity Systems (e.g., politics, banking)

**5.** Transcendental Systems (e.g., beyond knowledge or comprehension)

According to this classification, manufacturing systems are clearly a combination of Human Activity Systems and Physically Designed Systems which immediately identifies the human or social subsystem within manufacturing. Although this observation may seem trivial at this stage, we shall see later that many of the so-called system design tools and methods clearly avoid human aspects altogether. The issue here is not that these tools and methods are wrong, but rather that they can only form part of the total analysis of manufacturing systems.

An entirely different perspective on systems yields another interesting classification. Boulding has classified systems according to level of abstraction or complexity. Figure 1-2 illustrates the various levels of abstraction Boulding associates with GST. This hierarchy is often used in the explanation of GST and is an adaptation of Boulding's *hierarchy of systems*.[11] The first level in the hierarchy *frameworks* represents the most elementary of systems; for example, static structures (e.g., camshaft, skeleton, formal company organizations, rock) are usually not treated as systems. The second level *clockworks*, consists of those systems that use some sort of timing mechanism. Systems such as self-winding clocks and simulation tests are closed systems which constantly move but have no interaction with their environment. *Cybernetics* brings the concept of feedback to elementary closed systems. Thermostats, for example, are clockworks with feedback. The feedback alters the performance of the thermostat. However, the feedback in this example does not alter the goals of the system.

The next level of system is the *open system*. At this level of system, elementary forms of life are introduced into the hierarchy allowing systems to interact with their environment in order to change their behavior. *Genetic-societal* systems are subsystems that specialize and exchange information with other subsystems. The *animal* system introduces concepts to the system such as mobility, self-awareness, and goal orientation. In this system the organization of the subsystems and their interaction and communication are increasingly complex. The *human* system adds a new dimension to that of the animal system—intelligence. Intelligence gives the human system the ability to think about the future, to think about its goals, and to think about how to reach these goals. A level above this is the *social organization*. Here human systems come together to form organizations that have their own combined goals, needs, and ways of achieving goals. The final level of Boulding's classification is the catch-all term *transcendental* for all other systems not yet comprehended by today's systems analysts.

In Boulding's classification, manufacturing systems clearly go all the way to the top. Except in rare instances of fully automated manufacturing subsystems (e.g., unmanned robotic work cell), the manufacturing systems can be said to reflect the behavior of the social organization. Like Checkland's, this classification serves to illustrate once again that the manufacturing system behaves as a complex system, with groups of interrelated and interdependent subsystems interacting to form a collective system for the attainment of common goals.

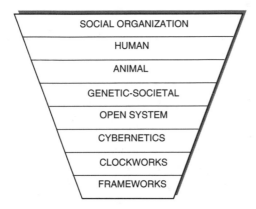

**FIGURE 1-2.** Levels of abstraction.

### 1.3.3 Systems Theory and Manufacturing Systems

The benefits of GST for the analysis of manufacturing systems are numerous. Many of these benefits are derived from the perspectives offered to designers on the behavior of systems rather than from any tool or technique. If this perspective can be shared by each of the designers of manufacturing systems, then perhaps many of the problems outlined in the earlier sections of this chapter can be eliminated. Systems are designed by different people. Unless a common way of looking at systems can be identified, systems designers will have serious problems in communicating with each other and achieving results. As Boulding puts it:

> "physicists talk to physicists, economists to economists—worse still, nuclear physicists to nuclear physicists. One wonders, sometimes, if science will not grind to a stop in an assemblage of walled in hermits, each mumbling a private language that only he can understand."[11]

GST is a way of combating these isolationist tendencies amongst engineers, software programmers, users and systems suppliers.

A second feature of GST as applied to manufacturing systems design is that it allows various subsystems to be treated together. In manufacturing systems there are numerous subsystems that are functionally divided as for example Product Design and Assembly. Subsystems can also be hierarchically and sociotechnically divided. In GST all parts of the system are analyzed together. The analysis is first used to develop a broad context for the problem which is then subjected to detailed analyses of the individual subsystems. The benefits and power of systems theory and, in particular, general systems theory are perhaps best illustrated by its contribution to state-of-the-art thinking in manufacturing systems design. This thinking has contributed many new concepts, techniques, and methodologies. These areas will be discussed later and include concepts such as learning organizations, sociotechnical design, and structured analysis. Also to be discussed are other areas that have evolved from this open systems theory are system design tools and methods, systems architectures and models, various standards, as well as a number of system planning tools.

## 1.4   MANUFACTURING SYSTEMS

The above discussion focused on the need for taking a systems approach to manufacturing systems design. The system which is of interest to this discussion is manufacturing. Within the manufacturing system many subsystems can be identified. Mentioned already are robotic work cells and human activity systems. But of course there are many other subsystems such as software systems, flexible manufacturing systems, management information systems, accounting systems, and the numerous *computer-aided* systems. Classification of manufacturing system types not unlike the classifications of systems by Checkland and Boulding is entirely a matter of perspective. Throughout this book a number of perspectives of manufacturing will be discussed largely depending on whether the discussion is based on social developments or technical developments.

Here two perspectives of manufacturing are presented that highlight two areas of discussion. The first perspective (functional) looks at manufacturing in terms of the functions it carries out. The main purpose of this view is to highlight the context for the functional aspects of manufacturing systems design. The second perspective (management) looks at various issues that surround the management of manufacturing systems, in particular the management of change. Issues such as company strategy and critical success factors are important here. Each of these perspectives is supported by general block diagram models or *frameworks*.

### 1.4.1 Manufacturing—A Functional Perspective

In the manufacturing environment a number of key generic functions may appear. In the Harrington model, for example 23 such functions are defined in the description of manufacturing activities.[18] From a different perspective the Scheer model of manufacturing identifies 21 functions (see Section 2.2.4 for both Harrington and Scheer models).[19] Figure 1-3 is a block diagram of the manufacturing enterprise, which identifies a number of generic functions and their interaction with each other from a manufacturing systems design perspective. The main functions concerned with the issues being discussed in this book are given separate boxes. The arrows between the functions indicate the interactions between various functions. Some functions are shown layered onto others to indicate the considerable interaction between these functions. The interaction between Manufacturing Systems Design and Operations is important and of particular interest to this discussion. A general introduction to each of the main functions is appropriate for clarification.

#### Operations

Operations is the function within manufacturing responsible for converting customer orders and raw materials into completed products, as shown in Figure 1-4. In this figure, Operations is shown as comprising a number of elements, interacting with each other to varying degrees. Operations can be seen to be positioned at the hub of manufacturing, lying at the crossroads between product information flow and order information flow. In this regard the design and integration of operations is crucial to the integration of the entire manufacturing enterprise.

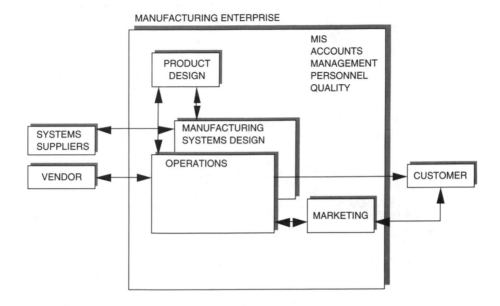

**FIGURE 1-3.**   Functions in manufacturing.

Within Operations, Process Planning involves the conversion of product information into manufacturing information such as process plans, shop-floor information, and material specifications. Operations Planning and Control is responsible for the planning of orders, materials, and available resources in order to meet customer demands. This function is also responsible for the control of these orders from order progress, through materials control, quality control, and on to maintenance.

Production represents the activities normally associated with the shop floor. It is the function of physically processing and assembling materials, moving and storing goods, providing information to the planning and control activities and executing in realtime the instructions and information provided by the planning and control activities. A useful classification of the five primary activities carried out in production is as follows:

1. Processing (machining, forming, casting, heat treatment, etc.)

2. Assembly (joining two or more components together)

3. Material Handling & Storage (moving and storing materials)

4. Inspection and Testing (quality control of components)

5. Control (maintenance, order, and resource management)

Each of these functions requires different technologies, which range from Computer Assisted NC Part Programming in Process Planning, MRPII in Planning and Control, and perhaps NC Machine Tools and Robotics in Production. Each technology needs to be developed and integrated in order to provide an integrated string of processes which together satisfy the goals of the company. The primary functions responsible for this development are Manufacturing Systems Design and Management Information Systems (MIS), which is shown as a separate function in Figure 1-4. It will be seen, in a later chapter, that the specialists normally associated with the MIS function become part of the man-ufacturing systems design function as soon as their expertise is required for a particular project and vice versa. Therefore, in the discussion of Manufacturing Systems Design which follows, the activities indicated for the larger projects involve both groups of devel-opment personnel.

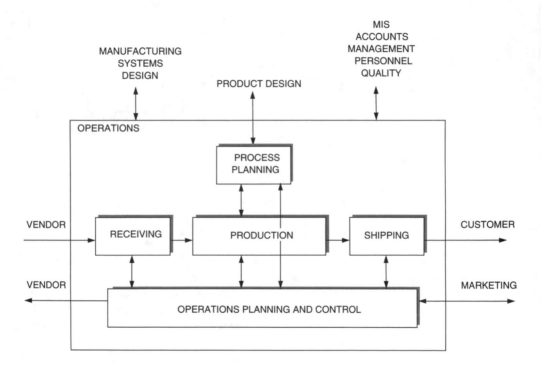

**FIGURE 1-4.**   Functions in operations.

## Manufacturing Systems Design

The Manufacturing Systems Design function is primarily responsible for the provi-sion of resources for the operations function. It is also responsible, in cooperation with the activities of MIS, to provide resources for Product Design and Marketing. The activities carried out by the function can be divided according to the relative scale of the project being undertaken.[5] There are four primary activities as illustrated in Figure 1-5: (1) Pro-vide New Core Process, (2) Provide Next Generation Process, (3) Provide Single Depart-ment Upgrades, and (4) Provide Tuning and Incremental Upgrades.

*Provide New Core Processes.*    This activity is usually associated with a major new product introduction or *green field* start-up. The process technology will be innovative, and the project as a whole will have a very high-profile status requiring a large amount of resources. The development of new core processes for the processing of a newly invented material is an example of an activity in this area. For example, in the development of laser readable compact disks (CDs), a new process was needed to produce the disks at the required high accuracy and volumes. The solution to the new process is innovative and not modeled on any older process.

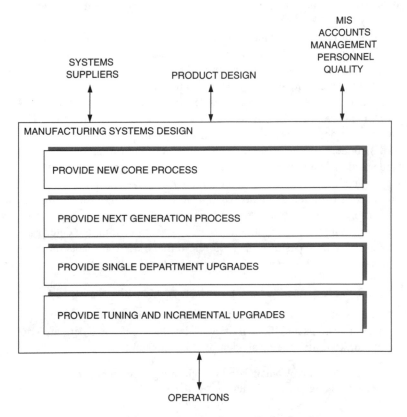

**FIGURE 1-5.** Manufacturing Systems Design functions.

***Provide Next Generation Processes.*** The provision of next generation processes involves updating existing processes to reflect state-of-the-art technology without changing the core process itself. As with the previous activity, projects in this classification require major resources. An example here would be the updating of an entire manufacturing facility to use state-of-the-art equipment, material flow concepts, and information systems. The core operations such as boring, turning, welding, and assembly will remain the same. However, the methods by which these operations are executed may radically change.

***Provide Single Department Upgrades.*** This activity involves upgrading single departments which may involve either new systems installation or the integration of existing systems. The implementation of a flexible manufacturing system incorporating a few CNC machines, material handling systems, and computer control would be a typical project in this activity.

*Provide System Tuning and Incremental Upgrades.*   The final activity in this definition involves the provision of incremental upgrades to existing manufacturing systems. The types of activities involved here could include the replacement of individual machines, fixtures, or redesigning a manual work station. Maintenance is also a major activity here.

With respect to manufacturing systems design and the need for new approaches to the integration of systems, the areas of primary interest are the projects being carried out in the provision of *next generation processes* and *single department upgrades*. While the other two areas—*new core processes* and *system tuning and incremental upgrades* are important, they open up other areas for discussion outside the scope of this book. Both of these areas can clearly benefit from a new development approach. However, the primary integration and development concerns will tend to lie in other areas such as innovative skills, understanding material characteristics, interpretation of long-term requirements, and specialist knowledge often generated through speculative research.

The development of a *next generation process* or *single department upgrade* involves a complex sequence of operations. It requires many organizational as well as operational changes to the system being developed, and because of its scope, it may also require changes in many other associated systems within the enterprise. The nature of this change and the types of tools required to make these projects more successful also mean that the manufacturing systems design function itself is also subject to change.

## Development and Operation of Manufacturing Systems

It is important to consider the differences between the activities of operating manufacturing systems and the activities of developing them. There are two parts to the design of manufacturing systems as identified by Chestnut:

> "the overall problem of systems engineering is composed of two parts one being the systems engineering associated with the way that the operating systems itself works and the other with the systematic process of performing the engineering and associated work in producing the operating system."[20]

As stated earlier, this book is concerned with the activity of manufacturing systems design or performing the engineering and associated work in producing operating systems. It is not concerned with any new operational concepts. In the operation of manufacturing systems, concepts such as *one-of-a-kind* production, production activity control, *engineering to order*, and *flexibility* are important. However, in the development of manufacturing systems other concepts such as design tools, project organization, systems theory, and sociotechnical design are important. Some design tools clearly incorporate both design and operation concepts. For example, group technology is clearly a design tool in that it helps in the analysis of material flows and process plans. It is also, however, an operational concept in that the results of the analysis are generally grouped machine cells. In the development of manufacturing systems, group technology is an exception. Almost all other development tools are not based on operational concepts. For example, a simulation tool only provides a method for modeling a system. The operational concepts must come from the designer. Figure 1-6 shows more clearly the difference between development and operational activities.

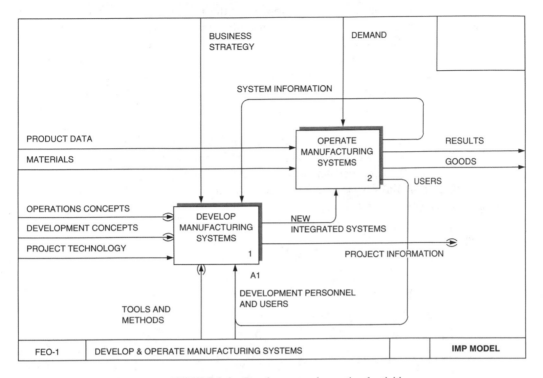

**FIGURE 1-6.** Development and operational activities.

From Figure 1-6 it is clear that manufacturing operations and manufacturing systems design have completely different sets of inputs and outputs. This distinction is important, because often designers confuse the two. For example, many project teams cite Robotics or MRP as goals. Robotics and MRP form part of specific operational paradigms and cannot be goals for systems design. Goals such as improved material flow, reduced lead time, and reduced inventory may lead to the use of Robotics and MRP, but only because these paradigms provide a solution. Another way of looking at this distinction is to note that projects mainly fail only because of the design process and not because of the operation paradigm. For example, if someone were to say that JIT failed in his or her particular manufacturing system, one should be quick to point out that it was not JIT that failed but rather an engineering process that perhaps incorrectly identified JIT as a solution to a particular problem. Therefore to successfully change the manufacturing environment we must concentrate on manufacturing systems design. A good manufacturing systems design approach will identify and implement a good operations paradigm. The reverse is seldom true.

## 1.4.2 Manufacturing—A Management View

The massive amount of money currently being spent on new manufacturing systems demands a comprehensive and systematic approach to their development and maintenance. A strategy needs to be developed for the manufacturing system in order that new investments adhere to some predefined objectives which will guide the growth and direction of a manufacturing system toward desired objectives.

The manufacturing strategy provides this approach, and its proper development can offer real competitive advantages for a company. The development of a manufacturing strategy requires information on a wide range of topics not the least of which are the overall business philosophy, business strategy, and other functional strategies. Figure 1-7 shows the relationship developed between the manufacturing strategy and other strategies present in a company.[21] In general, each strategy must support the business strategy which in turn takes cognizance of each functional strategy and functional capability. An interpretation of each of these elements is seen as appropriate here in order to create an understanding of the links between the traditional approach to manufacturing strategy and the traits of general systems theory.

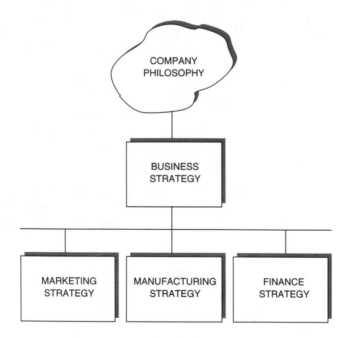

**FIGURE 1-7.**   Different strategies in a company.

## Company Philosophy

The effective implementation of a manufacturing strategy normally takes a number of years of perseverance on the part of manufacturing management. The necessary environment for this perseverance must first be created, and this is normally described as the company philosophy, which can be defined as:

> "the set of guiding principles, driving forces, and ingrained attitudes that help communicate goals, plans, and policies to all employees and that are reinforced through conscious and subconscious behavior at all levels of the organization."[22]

In short, it is the common set of values that all employees share at some time during their participation at work. The following views illustrate a simplified example of company philosophy.

1. People are the key resource.

2. Customers are never wrong.

3. Best quality product is produced.

4. State-of-the-art technology is used.

5. Employees work together toward a common goal.

Such philosophies have a direct impact on the strategies developed by a company and in particular on the manufacturing strategy. A properly managed manufacturing function plays a significant role in defining, supporting, and enhancing a company philosophy. This in turn helps to create competitive success in concert with all the other functions of the business.

## Business Strategies

The business strategy of a company is primarily concerned with the broader issues that concern the business as a whole. The main factors involved include (1) the identity of the business within its competitive context; (2) its business goals (3) the methods of competition; (4) the contribution of each product and function to the goals of the business; and (5) the allocation of resources among each product and function. The competitive context of a business determines its performance in areas such as innovations, cohesive culture, and implementations.[23] Two major factors underlie the choice of business strategy. The first is long-term profitability and the issues that help to achieve this. The second is relative competitive position. A manufacturing firm may be in an attractive business but because of its inferior competitive position it may not have long-term survival prospects. To sustain long-term profitability it is important to develop a business strategy that maintains a competitive advantage while management at the same time seeks out new business opportunities.

Business strategy is the parent or more general description of each of the functional strategies. The aim of functional strategies is to execute the business strategy through various tasks and programs. Figure 1-8 illustrates a general framework within which the business strategy and overall business plans are created. In this framework, which can equally be applied to strategies and plans at other levels in the business hierarchy, seven levels are identified as being part of the planning process.

Increased long-term profit is generally perceived as one of the main missions of a manufacturing company. However, the way in which this profit is produced dictates the various types of activities that the company performs. The objectives or goals of a manufacturing business are the ends toward which all activities are aimed. One clear objective is to produce or increase profit. The term *profit* needs to be properly defined, if it is going to be the only objective by which a manufacturing business is to be measured. The business strategy of a manufacturing company is the determination of the basic long-term objectives and the adoption of courses of action and allocation of resources necessary to achieve these objectives.

**FIGURE 1-8.**   Stages in business strategy.

Policies are the general statements or understandings that guide or channel thinking and action in decision making. These may be in the form of written documents or simply implied in the actions of managers. Policies in general define an area within which a decision is to be made and ensure that the decision will be consistent with, and contribute to, an objective. Policy is a means of encouraging discretion and initiative, but within limits depending on the individual concerned and his or her position in the management hierarchy. Procedures are guides to the action of carrying out various activities, and they detail the exact manner in which these activities must be accomplished. Procedures usually follow a chronological sequence and are documented formally in some type of manual. Rules are the simplest type of plan to interpret. They state clearly and unambiguously what can and cannot be done as a result of the strategies, plans, and procedures laid out in earlier stages.

Programs are complex associations of goals, policies, rules, task assignments, steps to be taken, resources to be employed, as well as other elements necessary to carry out a given course of action. Programs seldom stand alone and are normally part of a number of programs which interact with each other to varying degrees. Programs require budgets in order to be carried out. Budgets are statements of expected results from the various programs expressed in numerical terms.

The activity of creating a business strategy and all of the plans and policies that it implies is perhaps the most basic management activity. The end result of this activity is the creation of more detailed strategies in the areas of marketing, manufacturing, and finance as well as any other strategically important area. Within the context of the technology being discussed in this book, understanding the influence of the business strategy on the more detailed manufacturing strategy is important. However, as will be discussed later, the impact of other functional strategies on the manufacturing strategy can be just as important. In this regard the effect of the marketing strategy is perhaps the most important, since this impacts on manufacturing issues such as capacity, lead time, quality, and other customer-related factors.

## Manufacturing Strategy

Manufacturing strategy is primarily concerned with issues surrounding the manufacturing systems design function as described in an earlier section. It has been described as having five key characteristics which are:[22]

1. Time Horizon
2. Impact
3. Concentration of Effort
4. Pattern of Decisions
5. Pervasiveness

Manufacturing strategy is usually concerned with a finite time horizon, typically three to five years. Within this time horizon, various policies, developments, and results will all contribute to implementing the strategy. The impact of the strategy will over time (perhaps not until the end of the five years) show the relative success of the strategy through a number of critical success factors. Once identified, these factors will require a concentration of effort which will in effect place *blinkers* on developers so that only those factors that are important are addressed through the various policies and developments. Each of these policies and developments will require a varying degree of decision making in order to effect their realization. Decisions must stem from the overall manufacturing strategy and be compatible with each other so that they follow a consistent pattern. Finally, because the strategy will essentially embrace a wide spectrum of activities, from major systems development to day-to-day operations, all levels of the organization must act consciously and instinctively to reinforce that strategy and help it to become a success.

## Critical Success Factors

The development and implementation of a manufacturing strategy requires an understanding of the culture and philosophy of a company. This understanding is important in order to set the ground rules for the manufacturing strategy. In order to develop it further, knowledge must be gained of the existing manufacturing system and the factors that are important in quantifying its performance. Various critical success factors can be used to express this knowledge, and it is these same factors that can eventually be used to measure the effectiveness of the manufacturing strategy. Figure 1-9 gives ten typical factors that are used in the development of a manufacturing strategy. These factors have been derived from a number of sources.[5, 24, 25]

```
1. CAPACITY
2. QUALITY
3. TECHNOLOGY
4. INTEGRATION
5. HUMAN SYSTEM
6. FLEXIBILITY
7. PRODUCTIVITY
8. FACILITIES
9. PLANNING & CONTROL
10. TIME
```

**FIGURE 1-9.**   Ten critical success factors.

Capacity is perhaps one of the more frequent reasons for changes to an existing manufacturing environment. Capacity issues usually revolve around decisions regarding capacity increases—at what time and in which functional areas? In the last decade, quality became a key factor in the manufacturing strategy of most industries. Increased customer awareness and global competition have made the production of high-quality goods and systems imperative for the success of manufacturing companies. Quality issues have usually been concerned with product-component quality, but the scope has now been increased widely to include the quality of machines, systems, procedures, personnel and practices. Technology factors concern such issues as: Which technology should be used for a given process? How should this technology be integrated with other production stages? What should be the degree of automation or human-centeredness of the new system?

Integration is sometimes regarded as one of the new critical success factors that many companies use to justify heavy investment in computerization. As mentioned earlier, the term *integration* can in fact be viewed from a wider perspective and is discussed in more detail in the section that follows. In recent years the recognition of human systems as an essential part of manufacturing has led to the human factor being essentially reintroduced into the vocabulary of manufacturing strategists. Traditionally associated with this parameter are factors such as skill level, reward systems, and security. However, new factors such as increased operator responsibility and accountability, and increased operator participation are finding their way into long-term manufacturing strategies.

Getting a quality product to the customer is only one side of the issue for increased market share. Another side is getting the product to them on time and with the customization that they demand. Two types of flexibility are popular. Product Flexibility with factors such as timeliness of design change requests and Order Flexibility with factors such as lead time. Productivity will always be a key factor in manufacturing strategy. Increasing productivity not only leads to increased profits for the company but also increases the standard of living for the whole community. Facilities factors are concerned with the provision, expansion, location, specialization, and maintenance of existing and new facilities. Planning and Control is concerned with planning and control functions within the manufacturing environment. Issues such as materials control, sourcing, value-engineering, make versus buy, procedures, and planning concepts are all examples of decision areas to be addressed by Planning and Control. The final factor, Time, reflects a company's ability to design, manufacture, and deliver products to the marketplace in the shortest possible time. While Time is inherently a part of such factors as Integration and Productivity, its separation here into a unique factor reflects its growing criticality in recent years to the success of a company's competitive position.

Each of these critical success factors is important for the creation of specific goals or objectives for the company. Goals are more project specific. Typical goals in the context of manufacturing systems design are (1) reduction of inventory; (2) improvement of productivity; (3) improvement of quality; and (4) reduction of manufacturing lead time. While all mentioned factors essentially carry equal weight, one factor, Integration, has stood out more than all of the others in recent years, as a possible key to success for manufacturing change. This may be a trend, or it may simply reflect the current state of technology. There is little doubt however that Integration is now seen by most western companies as the primary factor for positive change in the manufacturing environment.

Many of the parameters described above are very closely interrelated and may share common goals for the manufacturing strategy. For example, the introduction of quality circles may improve the quality of products but it may also improve the working environment of the personnel working within the enterprise. Decisions regarding each of the factors discussed go to make up a particular company's manufacturing strategy. These decisions will be influenced by the business philosophy and business strategy of a particular company. The overall result of these decisions will be the execution of various programs which in turn will provide tasks and projects for implementation on the shop floor.

### Approach to Project Development

The development of a manufacturing strategy creates the foundation for a comprehensive plan for the deployment of resources. It also gives an indication of the time scale and expectations over a three-to-five year time horizon. The effect of this strategy as it flows down in the hierarchy is to explode into a large number of plans and even larger number of tasks and projects. Hence, each task and project and each expenditure and change in the manufacturing system are direct results of this strategy, which in turn can be said to be a direct result of the business strategy and ultimately of the business philosophy of the company concerned. This explosion of information in a controlled fashion creates the comprehensive and systematic approach to systems development deemed necessary for successful manufacturing competitiveness.

As each of these actions, whether they be individual projects, tasks, or procedures, becomes implemented, they create a back flow of control information which through careful management implodes to close the loop on the manufacturing strategy's effectiveness. Also, specific projects cannot always be planned from a top-down approach. Some projects *bubble up* from the shop floor, creating requirements for funding and resources. This creates a bottom-up flow of requirements' information which must be matched and married with the various strategies or discarded. This closing of the information loop or feedback from the tasks and projects facilitates effective management and control of the manufacturing strategy and leads to continuous improvement of the whole systems development process. This basic effect of the implementation and monitoring of the manufacturing strategy is sometimes described in terms of top down planning and bottom up implementation as illustrated in Figure 1-10.

### Setting the Stage for Integration

The issue of integration has already been discussed in terms of it being one of the key factors in manufacturing strategy. Over the past few years, improvements in other factors such as productivity, flexibility, and quality have only been possible in some companies when using the integration factor as one of the key project parameters. In this book integration is seen as a key factor in the progress of manufacturing companies in the coming decade. No matter how companies improve in terms of other factors, they will sooner or later reach their upper limits. Integration alone will offer the continued improvement necessary for companies to remain competitive.

TOP DOWN
PLANNING

BOTTOM UP
IMPLEMENTATION

STRATEGY

PLAN

IMPLEMENTATION

**FIGURE 1-10.**   Approach to Strategy Planning and Implementation.

## 1.5   INTEGRATION

Computer Integrated Manufacturing (CIM) is undoubtedly one of the most influential concepts to have been introduced into the manufacturing world in recent times. It has been the basis of many major public and privately funded research projects and has excited the interest of most companies throughout the developed world. Yet the concept itself is ambiguous and lacks explicit definition. Perhaps this is one of the reasons why some companies can now go either hot or cold when trying to address any one of the many issues that have been associated with CIM over the years. In general, the term Manufacturing as used in CIM explains the context for CIM development; the term Integration represents the key goal of CIM; and the term Computer represents one of the most popular mechanisms used in achieving the key goal. In this simplified explanation it is easy to immediately drop the terms Computer and Manufacturing and concentrate instead on Integration and what it means.

A number of approaches for defining the term integration which is less specific than the term CIM have been documented. Konig offers one approach for defining two specific types of integration in the manufacturing firm.[26] The first is the integration of the functions in the product cycle (traditionally CAD/CAM). These functions include Design, Process Planning, and Manufacturing. The second is the integration of the order cycle. Functions involved here include Customer Requirements, Marketing, Manufacturing Planning, Manufacturing, Shipping/Receiving, and Suppliers. Both integration types meet at manufacturing. This purely functional view of integration is extended by Maniot and Waterlow who propose another three types of integration in addition to functional integration.[27] Their second integration type is technical integration where they visualize integration through computer communications. Their third type is information where they isolate all information flows in manufacturing for integration. Finally, their fourth type is strategy integration, in which they view the need to have various company strategies integrated in order to form a common set of company goals.

Another classification on the same theme comes from Voss who describes five types of integration: Strategy, Material Flow, Technical, Information, (and the first hint of the human subsystem) Organization.[28] Rockwell proposes somewhat similar integration types: Logistics, Management, Facilities, Information, Culture, and Strategy.[29]

One classification of integration types that has been gaining in popularity over the years is given in CIM-OSA. In a prenormative standard CIM-OSA generally defines four types of integration: Function, Information, Resource, and Organization.[30] Study of the standard reveals a classification that is perhaps less detailed than one would hope for and certainly one that requires further information. It does however have the advantage over the other classifications in that it is recognized as a prenormative standard, and in this respect it is a useful basis upon which to start. CIM-OSA is discussed in greater detail in Chapter 2.

In terms of manufacturing systems design there is a need to isolate only those integration issues associated with the design rather than the operation of systems. In this respect, a new perspective on Integration will be presented. It is defined in order to throw some light on the major issues that need to be addressed from a systems-design perspective. The new perspective uses the views of some of the above classifications but enhances them to isolate important issues for manufacturing systems design.

## 1.5.1 Types of Integration

One approach to the definition of Integration from the manufacturing systems design perspective is to look on integration as consisting of social as well as technical elements.[31] Social integration, as the name implies, involves the integration of people, their ideas, and the company's decision-making processes. Technical Integration is more concerned with the integration of technical subsystems, which by this definition include equipment, techniques, and procedures. Figure 1-11 shows a model of Integration which illustrates the following six individual types of integration operating in the integrated manufacturing environment: (1) Information Integration, (2) Data Integration, (3) Equipment Integration, (4) Management Integration, (5) Systems Designer Integration, and (6) User Integration.

### Information Integration

At issue in Information Integration is the nature of the data flowing through the system, or not flowing as the case may be. Another issue is whether this information is the best available in the system to exercise control over the manufacturing process and contribute to effective management. Ideally this information is that which is required for other systems to function productively. Various types of information will exist and can generally be categorized into planning information, control information, and historical information. It will include information on parameters such as materials, order status, quality levels, and so on. To quantify this information it is first necessary to analyze information requirements from the various systems in manufacturing, as well as from managers and users.

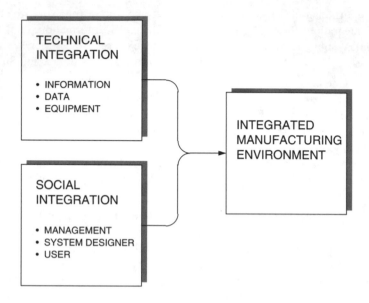

**FIGURE 1-11.**    Six types of integration.

These information requirements must then match with *obtainable data*. For example, in order to measure the productivity of a certain system, a manager may require certain data from the shop floor. These data may be difficult to capture and indeed if captured may require an excessive amount of effort to reformat. In this case, the information requirements should be reexamined so that the best match between obtainable data and management requirements is met.

### Data Integration

Data integration is concerned with the communication between various subsystems. This communication can be either electronic or manual. This area is currently typified by various standards initiatives such as MAP, TOP, and IGES. For most companies, however, the integration is not within their control or scope, and instead they must choose from the many vendors supplying data-integration solutions. One way for companies to address data integration is to improve procedures and policies to enable a more efficient flow of information. An important point here is that the solutions to this type of integration must be preceded by an analysis of information requirements. Otherwise much effort and expense can be wasted on systems that do not facilitate effective manufacturing control. For example, in order to capture data from a particular system comprising a number of CNC machines, it may be necessary to use a computer network and applications software that feed the appropriate data into a relational database. One of the main issues that may need to be addressed will be the communication protocols between the host system and the individual machines. This issue will raise a number of questions: Are the protocol specifications available for each of the machines in the system? If available, is modification and coding of the application on the host cost-effective? Is a database the most appropriate storage medium with respect to integration and access by the application program?

### Equipment Integration

Equipment integration is concerned with the physical integration of equipment within the manufacturing environment. At present, it is typically tackled through customization of the equipment and through various standards that are available or are under development. It is often an expensive activity where vendors do not adhere to standards. For example, for the CNC machines mentioned above, there will be a need for material movement between the machines. This can be implemented by conveyor system, automated guided vehicle (AGV) system, manual handling, or automated storage-and-retrieval system. Each of these systems will place major constraints on the layout and in some cases the customization of other equipment. An AGV docking system at one of the CNC machines may have to be customized. It can be a major expense with regard to docking sensors, hard-wire communications, software communications, and physical fixturing. Customization of the CNC, AGV system, and the supervisory software may be required.

### Management Integration

Management integration involves the integration of managers, their decisions, and their individual functional strategies in order to adhere to the business strategy and lead to effective problem solving and manufacturing control. It also involves study of the effects of a changing environment on the roles of these managers. This type of integration is perhaps the most difficult to define in the context of new system design. One might presume that there is good integration if a company is making profits and identifies the need for continuous change in the manufacturing environment. However, one area where this type of integration could be tested is in the formation of multidisciplinary project teams. Good integration may be said to exist where project teams can be formed that support the right level of loyalty to the group from the project members, while at the same time satisfying functional duties to functional managers. Similarly, poor integration may be said to exist when managers will not allow their people present in the project group to display loyalty to anyone other than themselves.

Another aspect of management integration involves the job functions of management. Evidence exists that the role of management and in particular supervisors will change dramatically as new systems are placed into the manufacturing environment. Some of these changes will include decreased need for monitoring and supervising, more time and need for coordinating with other personnel, more time investigating complex production problems, and a decreased ability to use traditional methods to motivate and judge operator performance. In short, managers and supervisors will need to spend the time saved by delegating responsibility to subordinates, by communicating with each other more effectively, and by providing a management team approach to problem solving.

## Systems Designer Integration

Systems Designer Integration recognizes that systems are designed by a number of functional groups (Management Information Systems, Manufacturing Systems Design, Production, Quality, Operators, etc.). Systems designer integration also recognizes the need to have these functional groups integrated into a project group to facilitate efficient improvement of systems. In the case of project teams, much of the control lies with the group leader. If a good leader is selected, much of the energy used in the rivalry between functions can be transformed into making the group more loyal to the goals of the project. This will be discussed in more detail later. Systems designer integration also recognizes changes in various job functions. These are the job functions in skilled trades, quality and production control, programming, engineering, systems analysis, accounts, and marketing. In the skilled trades, maintenance personnel will find their technological scope increasing dramatically with the implementation of new technologies. Personnel in quality and production control will shift their attention to quality coordination and master scheduling as operators take on more responsibility for inspection and daily scheduling. Programmers for CNC and other equipment will, through the use of more sophisticated applications, find their jobs becoming more systems oriented. Manufacturing engineers, design engineers, and systems analysts will find themselves working together in product teams rather than as functional entities within the manufacturing environment. Even accounting functions will experience change. New methods for cost-justification and performance measurement will need to be devised, and accountants will take a more active role in the design and operation of manufacturing systems. Marketing functions will also change. Marketing personnel will get more involved with the designer, manufacturer, and customer as an integrated team. Continuing on the concept of product groups, marketing will complete the cycle from customer to supplier.

## User Integration

This type of integration involves the integration of users within their social and technical environment. The primary issues here are ensuring that new systems support or augment the role of users to make their tasks more enriching and more productive. A number of key areas are covered here, from involvement in systems design to the development of job design and the man/machine interface. User Integration also involves the change in roles of users. Users in the future will take more control over their job from quality, scheduling, and design perspectives. Perhaps one of the key issues will be that of job design. In the past as well as the present, much effort in job design has been based on the scientific management approach. While many companies still use the technique, there is a rapid movement toward alternative methods. Other techniques such as Job Enrichment, Fault Detection, Sociotechnical Design, and Synergism are all finding increased application to varying levels in the modern manufacturing environment.

Each of these integration types can be combined to create the integrated manufacturing environment. Various issues can be identified when viewing only a few of the six types. System design projects may place more emphasis on one or more of these types of integration than others, depending on the perspective of the project group. In general, however, projects should take cognizance of all six to provide a truly holistic approach to integration issues.

## 1.6    SYSTEM MODELING

In the depression year of 1929, the first factory modeling techniques were used for planning and designing manufacturing systems. Models were used to illustrate the ways factories were to be laid out and material moved. In the present environment numerous models and modeling techniques abound to help designers describe and specify the manufacturing environment. Indeed, each new book published on the subject of manufacturing seems to put forward a new type of model or modeling technique. These range from the ambiguous yet simple block diagram to the sophistication of computer simulations and data flow diagrams. In general each model must be viewed from the perspective of its author. Despite the variety and complexity of modeling processes, modeling is and should remain one of the more powerful communication tools in systems design.

A carefully selected modeling methodology can have a major impact on the formulation of ideas and analysis of system variables. A good modeling environment can significantly improve all stages of project development, from auditing of existing facilities through financial analysis, system specification, system development, and on to acceptance and auditing of the final system. In this book it will be shown that different modeling techniques are appropriate to different stages and functions of manufacturing systems design. However, in putting across the ideas contained in this book two modeling approaches are adopted. The first is the more common block-diagram approach which is used extensively throughout the first four chapters. Block diagrams, which generally lack detailed information, are useful for supporting discussions in the text.

The second modeling technique used in this book allows for more detail and less ambiguity of information and information relationships. This technique, used extensively in Chapters 5 and 6, is called IDEFo. Since IDEFo is a formal modeling technique, a brief introduction to its principles and techniques is appropriate at this point. Indeed, for the reader to fully appreciate many of the ideas and much of the information presented in the chapters that follow, a general understanding of how to read IDEFo diagrams is essential. Following the next section the reader is invited to reexamine Figure 1-6 (which follows the guidelines of IDEFo) and perhaps to understand the information contained in the figure in its entirety.

### 1.6.1  IDEFo

ICAM definition zero, or IDEFo as it is more popularly known, is an activity modeling method for modeling activities and flows in a system. The IDEFo model consists of three components: (1) a set of hierarchically decomposed diagrams, (2) an accompanying text for the diagrams, and (3) a glossary of terms used in the diagrams.

IDEFo was developed during the Integrated Computer Aided Manufacturing (ICAM) program. In this program manufacturing industries who supplied the U.S. Air Force were to be given access to state-of-the-art factory models that would help them to understand and ultimately redesign their own factories. Before these models were created a baseline communication system was required in order that systems could be planned, developed, and implemented in the various aerospace companies. This baseline was called the Architecture of Manufacturing, and it required a language to express and document unambiguously many of its various concepts. One of the languages chosen was the subset of a tool developed at Softech called the Structured Analysis and Design Technique (SADT) by Douglas Ross.[32] This subset was called ICAM Definition Zero, or more popularly IDEFo. It was one of three modeling languages specified, the other two being IDEF1x for data-flow modeling and IDEF2 for dynamic simulation modeling. IDEFo can be generally described in terms of three basic concepts: Cell Modeling Graphics, Hierarchical Decomposition, and Disciplined Teamwork.

### Cell Modeling Graphics

IDEFo can be used to model the activities and the flows in a system through the use of rectangular boxes and arrows. Boxes represent activities, and arrows the input and output flows. Because they are activities, the text inside the box must describe, using an imperative, the precise activity which is being modeled (e.g., Make goods, *Implement* program, *Operate* machine). The way in which the arrows enter the box is important. Arrows entering from the top face are controlling inputs. These inputs control the execution of the activity (e.g., a *schedule* controls the work done in a machine cell). Arrows entering from the left-hand face are simple inputs; they do not control the activity but are used by it to be transformed into the outputs (e.g., *raw material* input is transformed into *goods*). Finally, arrows entering from the bottom face represent resources required by the activity (e.g., a machine is a required resource to produce a part from raw materials). Figure 1-12 illustrates these types of flows as they exist between two activities.

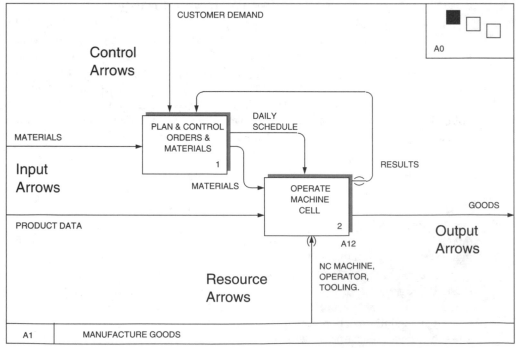

**FIGURE 1-12.**   IDEFo arrow conventions.

Also shown in this figure are activity numbers, an implied sequence of activities (diagonally from left to right) and feedback illustrating the time independence in the modeling method. Note that only flows that are important to the design project need to be shown, leaving trivial or less important flows out of the model. For example, power is an important resource for a machine cell. However, if the design project does not involve the analysis of power consumption, then resource flows about power, energy, or fuel can be left out of the model.

### Hierarchical Decomposition

Another powerful approach that is used in IDEFo is the stepwise revelation of detail through hierarchical decomposition. Each activity in a model represents a number of sub-activities, and each arrow represents a number of sub-arrows. So, for example, the activity of Operate Machine Cell can be viewed as comprising two subactivities—Manage Cell Operation and Produce Goods. In addition, input and output flows around the Operate Machine Cell activity emanate from, or are decomposed into, these subactivities. The concept of hierarchical decomposition is illustrated in Figure 1-13.

There are a number of rules regarding decomposition that must be observed if the true unambiguous nature of the model is to be maintained.

**1.** Each activity (or group of subactivities) is given a node number. In Figure 1-12 the activity Manufacture Goods has the node number A1. *A* represents the model and *1* the relative position of the activity in the model. The activity at the top of the hierarchy is normally denoted A0.

**2.** The node numbers of subactivities is determined by suffixing the node number of the activity to the subactivity number. In Figure 1-12 the second (No. 2) activity *Operate Machine Cell* has the node number A12.

**3.** Each activity box contains a title box which clearly illustrates the context for the activity (top right-hand corner), its activity title, and its node number.

**4.** Arrows that are not traceable to parent or child diagrams can be tunneled. Tunneled arrows appear in parentheses.

**5.** Activities that are further decomposed will have the full node number indicated outside the box on the bottom right-hand corner.

### Disciplined Teamwork

The final concept that will be discussed, and one that is often forgotten, is the social aspect of the IDEFo methodology. Procedures exist for developing and critiquing the models developed. The basic cornerstone of this procedure is the reader–author cycle illustrated in Figure 1-14.

In this cycle the author of the model is complemented by a reader whose task it is to check the validity of the model against the stated objectives of the model creation (e.g., *to model the material and information flow around a turning cell*) and the perspective to be adopted (e.g., *from the perspective of the production supervisor*). The basic mechanics of the cycle are simple. The author creates the model which comprises diagrams, supporting text, and a glossary. The reader then reads the model to check that it conforms to the stated objectives. In this way the unambiguous nature of the entire modeling procedure is maintained.

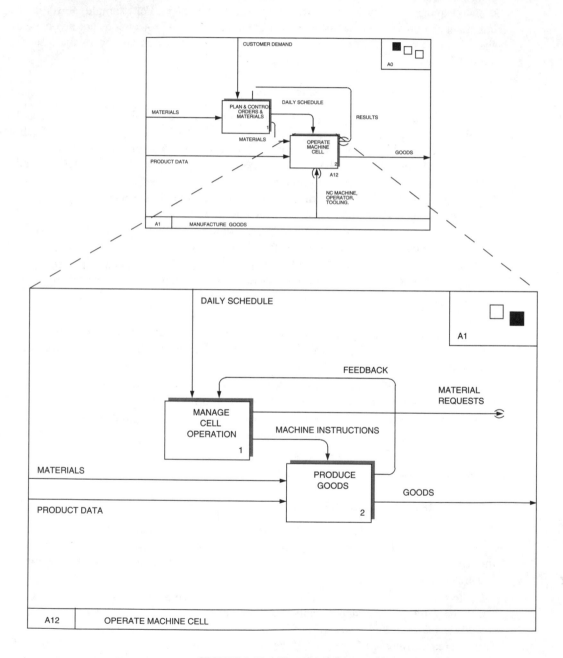

**FIGURE 1-13.**   Hierarchical decomposition.

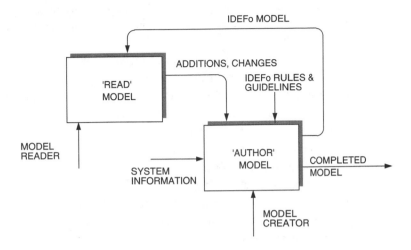

**FIGURE 1-14.**   Reader–Author cycle.

## Principles of IDEFo

The IDEFo methodology was developed to adhere to six guiding principles for systems modeling, each of which provides an explanation for the rigor and effort that needs to be expended when creating an IDEFo model. The six principles are:

1. Cell Modeling Graphic Representation: The *box-and-arrow* graphics of an IDEFo diagram show the manufacturing operation as the box and the interfaces to/from the operation as the arrows entering/leaving the box. In order to be able to express real-life manufacturing operations, boxes operate simultaneously with other boxes, with the interface arrows providing *constraints* as to when and how operations are triggered and controlled.

2. Conciseness: The documentation of a manufacturing architecture must be concise to permit encompassing all of the subject matter. The linear, verbose characteristics of ordinary language text is clearly insufficient. The two-dimensional form provided by a blueprintlike language has the desired conciseness without losing the ability to express relationships such as interfaces, feedback, and error paths.

3. Communication: There are several IDEFo concepts that are designed to enhance communications:

   a. Diagrams based on simple box and arrow graphics

   b. English textual labels and glossary

   c. Gradual exposition of detail

   d. Node chart for providing quick index of diagrams

   e. Limit of not more than six boxes per figure

**4.** Rigor and Precision: The rules of IDEFo require sufficient rigor and precision to satisfy ICAM architecture needs without overly constraining the analyst. IDEFo rules include:

   a.  Detail exposition control (no more than six boxes)

   b.  Limited context (no omissions or unnecessary detail)

   c.  Diagram interface connectivity (through node and box numbers)

   d.  Data structure connectivity (in parentheses)

   e.  Uniqueness of labels and titles

   f.  Syntax rules for graphics (boxes and arrows)

   g.  Data arrow branch constraints

   h.  Inputs are separate from controls

   i.  Data arrow labeling requirements (minimum labeling)

   j.  Minimum control of function (all activities controlled)

   k.  Purpose and viewpoint for the model

**5.** Methodology: Step-by-step procedures are provided for modeling, review, and integration tasks. In the case of public domain projects, the IDEFo manual is available, and private courses can be arranged by numerous education and consulting institutions.

**6.** Organization versus Function: The separation of organization from function is included in the purpose of the model, and carried out by the selection of functions and interface names during model development. This concept is supported in the modeling methodology and ensures that organizational viewpoints are avoided. This principle is critical in the design approach employed by designers when using IDEFo. Real change can only come about when the design approach is insulated against existing organizational constraints.

## Approach to IDEFo Modeling

In the development of individual factory models by internal work groups, there can be various approaches to the application of the IDEFo modeling methodology. In general, the approach must adhere to the principles and techniques outlined above. The following is a briefly summarized six-stage approach that has proven to be effective:

**1.** Select a viewpoint and a purpose for the model.

**2.** Limit the subject matter.

**3.** Create the context or top-level diagram (A-0, one box only).

**4.** Create top diagram in the model (A0, two to six boxes).

**5.** Create subsequent diagrams, text, and glossary.

**6.** Review material and check for purpose and viewpoint.

Additional pointers for the modeling procedure are:

**1.** Avoid trivial activities and flows.

**2.** Limit the necessary detail at each level.

3. Group related arrows and activities to simplify detail.

4. Be clear, precise, and consistent.

5. Think *control* and not *flow*.

6. Delay the addition of detail.

7. If in doubt, incoming flows should be a control.

8. Annotate as you develop each diagram.

The development of an IDEFo model will be a function of the author's perspective and purpose and the degree to which he or she adheres to the general guidelines laid down in the IDEFo methodology. There are of course many benefits to adhering to these guidelines, but in general some short cuts can be taken particularly with respect to the reader–author cycle and archiving of models (an issue not discussed above but also part of the IDEFo methodology).

IDEFo is a comprehensive systems development methodology for use in the functional description of systems. The technique itself is in widespread use in all phases of manufacturing, and is clearly simple yet rigorous enough to satisfy systems development issues. It does however have a limited application to the more conceptual levels of systems modeling. In order to gain increased definition that can facilitate the development of, for example, software systems, more specific systems development methodologies must be used, such as data-flow diagrams. These more specific methods complement the excellent design practices inherent in the IDEFo technique.


## 1.7   CONCLUSION

Setting the context for manufacturing systems design is important for our understanding of the change process in manufacturing systems. To understand manufacturing systems we must also understand the general concepts of systems, in particular, systems theory. Systems theory allows us to set a foundation for any further analysis of manufacturing and any approaches we may adopt in manufacturing systems design. The manufacturing systems design group is responsible for implementing change on or near the shop floor. In this respect manufacturing systems design is responsible for changes in the functions of operations planning and control, shipping, receiving, process planning, and production. The scope of manufacturing systems design is therefore established. It can only be increased by establishing broader design groups, including such functions as management information systems.

Company strategies must be the major prime movers for change. Two approaches reflect the state-of-the-art thinking in this area. The first is that all change must emanate from and be controlled by a manufacturing strategy. This is in effect a top-down process. In addition, change must also be accommodated by user requirements, which promote the bottom-up implementation approach. Both top-down and bottom-up approaches complement each other. Critical success factors for systems integration and measuring the way manufacturing strategies are implemented are important in *framing* the scope of integration projects. Various views of the term *integration* are possible, and in the end a new synthesized view comprising social as well as technical integrations is more suitable for exploring the issues surrounding manufacturing systems design.

In the two chapters that follow we shall look at the various state-of-the-art tools, methods, and techniques available for integrated systems design. These design systems include both technical and social planning approaches. The use and interpretation of both of these areas is in keeping with the holistic approach to systems, recognizing that both social and technical aspects of systems design are dependent on each other.

# Technical Planning Systems

## 2.1   INTRODUCTION

In the previous chapter we looked at a number of basic ideas about systems. This was done within the context of the activity of manufacturing systems design. We also saw that integration is now a critical factor in achieving the upper limits of system performance. In this chapter we shall focus on the tools and techniques that facilitate the technical side of integration. The importance of these tools and methods are identified by Joseph Harrington when he wrote: "When we understand our profession as a science, we can subject it to well known methods of analysis. Having analyzed, we can predict; we can determine what the basic parameters are and how they should be measured. When we can measure, we can control; and when we can control, we can succeed!"[18]

Harrington's work on manufacturing models and the use of some of these analysis methods and tools will be discussed later. This chapter also discusses other tools and methods and identifies a number of important areas used for technical integration. These areas have been divided into five sections: System Architectures and Models, System Modeling Tools and Methods, System Planning Tools, Standards, and Other Technical Planning Tools.

## 2.2   SYSTEM ARCHITECTURES AND MODELS

The development and implementation of manufacturing systems has always been the function of the manufacturing systems designer long before the concepts of CIM or integration ever became a key issue. But in the past ten years, with the increased focus on open systems concepts and integration, the question of the *approach* to systems development has become important. Within research on manufacturing systems a number of so-called generic architectures and models have been created using various systems concepts and modeling techniques. These architectures and models attempt to show us a view of generic and future factories and to provide such an approach for systems design. Before looking at some of these architectures it is necessary to first distinguish between the two terms—architectures and models. These two terms will be used repeatedly throughout this section.

## 2.2.1 Models and Architectures

The concept of a model is not new to us. Every child has at some time or other created a model of his/her favorite house either two dimensionally on paper or using three-dimensional equipment such as wooden blocks. The definition of a model is a little more formal: *A model is a qualitative or quantitative representation of a process or endeavor that shows the effects of those factors which are significant for the purposes being considered*. Formally, models can be defined as belonging to four categories. An *iconic model* is a representation of the form and/or properties of a subject (e.g., a wind tunnel for modeling the flow of air). An *analogic model* is a representation of the behavior of a subject (e.g., boolean logic diagrams for modeling the flow of current in integrated circuits). An *analytic model* is a representation of the mathematical or logical properties of a subject (e.g., $F=Mg^2$ models the force ($F$) due to gravity ($g$) pull on a body with mass $M$). Finally, a *conceptual model* is a representation of the symbolic behavior, form, logical, mathematical and/or properties of the subject (e.g., a flow diagram for modeling material flow).

In manufacturing systems design analytic and conceptual models have wide application. For example, analytic modeling is widely used in operation research which is typified by page after page of mathematical equations representing material flow in batch manufacturing. Conceptual modeling has always been used for sketching the layout of machines or flow of material. But only recently has it been developed to include modeling of such subjects as information. It is specifically this type of modeling (i.e., conceptual modeling of information flows) that this discussion is centrally concerned with.

Conceptual models are qualitative representations of the subject. In manufacturing they tend to be used primarily in the description of functions, activities, and flows. Unlike analytic models they have the potential of being very vague or ambiguous, which has both positive and negative effects. The positive effect is that they can be used in the representation of subjects which are themselves vague or *soft*. Initially, they can be used roughly for the generation of knowledge and then later redefined and remodeled to give a more accurate picture. For example, in the layout of machines, the designer may first sketch an initial design on paper for quick analysis, and then later work on the finer detail and eventual specifications. Similarly, the negative effect of conceptual models is that ambiguous or rough sketches can mislead other designers and in some cases lead to both information blocks (where two designers are talking about two different issues simultaneously) and error. In the models and architectures that are illustrated below, the conceptual modeling approach has been followed. The negative effects, however, have in many cases been overcome through the use of a structured approach.

Architectures, on the other hand, are structured plans upon which an organization or enterprise can be constructed and are defined in manufacturing terms as *a body of rules that define those systems features which directly affect the manufacturing environment into which the system is placed. These features include system configuration, component locations, interfaces between the system and its environment, and modes of operation.* Architectures contain both conceptual models as well as rules that help to translate the model into a working reality. Two types of architectures for manufacturing can be created, a reference architecture and a particular architecture. A reference architecture is a "first pass" architecture for a manufacturing system, that is to say it is very general and applicable to many types of manufacturing systems. A number of the architectures to be discussed later fall into this category. A particular architecture is one that can be derived from the reference architecture to suit a particular manufacturing organization. The only particular architectures to be discussed will be those presented in the case studies in Chapter 6.

## 2.2.1 Research on Manufacturing Architectures and Models

Research on manufacturing architectures can be divided into three categories: (1) research carried out through collaborative research projects, (2) research carried out by computer manufacturers, and (3) research carried out by individual researchers or research institutes.[33] Some of the more important architectures and models to be developed through these research initiatives are illustrated in Figure 2-1 (refer to the Glossary for an explanation of each term).

| SYSTEM ARCHITECTURES AND MODELS | | |
|---|---|---|
| Collaborative | Computer Manufacturer | Individual |
| CIM-OSA | IBM | HARRINGTON |
| CAM-I | DEC | THACKER |
| ICAM | HP | SCHEER |
| NBS | | RANKY |
| CALS | | YEOMANS |
| IMPACS | | |

**FIGURE 2-1.** Research on manufacturing architectures and models.

Due to the scale of investment the most significant work is carried out in the area of collaborative research. In the United States the CAM-I and ICAM programs are significant in this respect. In Europe the ESPRIT program, which has been in place since 1983, is continuing to advance the concept of precompetitive research for the benefit of industry through better standardization and insight into manufacturing operations. Collaborative research programs usually combine a mix of industry, university, and research institute interests. This collaboration, one can argue, provides the optimum use of knowledge and innovation from each of these three important sources, each of which will now be discussed.

## 2.2.2 Collaborative Research Architectures and Models

In this category five well-known architectures will be discussed separately. These are the ICAM, NBS, CIM-OSA, CAM-I, and IMPACS architectures. The perspective adopted in each of the discussions is one that yields insights into the development approach rather than the operations nature of the architecture or model. A short review of the CALS (Computer Aided Logistic Support) program is included at the end of this section. Strictly speaking, CALS is not an architecture but rather an ongoing array of solutions aimed at providing models and architectures for manufacturing.

### ICAM Architecture

During the late 1970s the U.S. Air Force identified a need for more integration to take place in order to facilitate better communication and design of manufacturing systems among its many suppliers of aircraft components. The USAF addressed the issue by setting up the Integrated Computer Aided Manufacturing (ICAM) project. The aim of ICAM was to create a so-called *Architecture of Manufacturing* which could then be used by industry in order to design or redesign subsystems within a manufacturing system. The architecture was created by a consortium consisting of members from manufacturing industries and private consultants. They used a number of modeling tools specifically designed or chosen so that the complex relationships between manufacturing activities could be modeled generically. Three tools were selected: (1) IDEFo for activity modeling (introduced in Section 1.6), (2) IDEF1x for data modeling based on entity-relationship diagramming, and (3) IDEF2 for event modeling based on state transition diagrams. The architecture recorded a common understanding of the manufacturing process by the members of the consortium. Details of the activities of manufacturing, the flows among activities, and the primary controls were established. Once detailed, the architecture could then be used by the members of the consortium and other manufacturers as a basis for the design of CIM subsystems.

The approach to the development of the Architecture of Manufacturing by the ICAM consortium was to use the modeling methodologies mentioned above (IDEFo, IDEF1x, and IDEF2) as well as a sound perspective on the *systems* nature of manufacturing. Three aspects of the manufacturing systems were addressed. The first aspect was *Context*, which identifies the subject matter of the model by describing its boundaries. The second aspect involved *Emphasis*, which was to be given to each of the potentially large number of activities contained in the context. Finally, the third aspect was to identify the *Purpose* for which the model was being created. This in turn determines its scope, depth, and ultimately its structure.

In order to create a generic architecture of manufacturing, ICAM began by looking at various individual *Factory View Models* of manufacturing from individual plants. These views are company dependant and concerned with the organization and structure of individual plants. Once established, individual factory views are incorporated into the *Composite View Model*, which is based on the essential decisions and activities needed to produce a product. The composite view is primarily oriented toward more conceptual and high-level activities within the manufacturing enterprise. The composite view also forms the baseline for the development of the *Generic View Model*, which is the synthesis of information created in the composite view to represent distinct functions of manufacturing. These functions represent the building blocks from which many other views can be formed during individual manufacturing systems design projects. The relationship between each of these three views is illustrated in Figure 2-2.

ICAM is an important contribution to the area of manufacturing research, if only because it was the first research program to identify the need for a common specification language among manufacturers (the IDEF language). Although IDEF is now being replaced by more customized sets of modeling tools, it set a trend for many of the research programs that follow. The architectures generated in ICAM are not generally available publicly—access is restricted. The benefits of the ICAM architectures, as with many of the others to be discussed, lie mainly in their contribution to standardization and the knowledge gleaned by readers on the approach adopted by their authors.

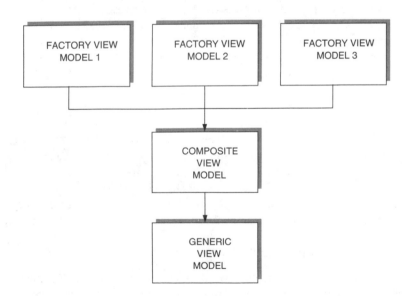

**FIGURE 2-2.**   ICAM architecture viewpoint.

## NBS Architecture

To promote standards development and the transfer of technology to U.S. industry, the U.S. National Bureau of Standards (NBS) established in 1986 an experimental test-bed known as the Automated Manufacturing Research Facility (AMRF).[34] The aims of the test-bed are essentially to develop systems with a greater degree of flexibility and modularity than existing manufacturing systems. In order to achieve these aims, a generic architecture called the Hierarchical Control Model for automated manufacturing systems was developed.

The approach to the development of the model was preceded by the development of a design philosophy. Three statements sum up the approach:

**1.** Partition of the systems into a hierarchy in which the control processes are isolated by function and communicate via standard interfaces.

**2.** System design capable of responding in real time to performance data derived from machines equipped with sensors.

**3.** Implementation with a distributed environment using recent advances in software engineering, microcomputers, and artificial intelligence programming techniques.

The NBS believed that one major reason why manufacturing systems vendors are not yet developing compatible products for CIM is that commonly accepted guidelines or specifications for factory automation modules do not exist. There is a need for a standard architecture that would allow users to build systems in increments and to buy these increments from competing vendors. A standard factory architecture must address all of the necessary functional, control, data flow, and interface issues.

Furthermore, the architecture should be based on fundamental scientific principles and be partitioned into submodules that can be readily understood by systems developers. For such an architecture, commands should be given in a top-down fashion with a successive breakdown at each level. The decomposition should be based on procedures, functions, or rules providing a string of lower-level commands. Each level should have a unique control language, developed for the particular level of abstraction and needs of the level. The NBS architecture was developed to consist of five levels of hierarchy, with each level breaking down into more activities. This is illustrated in Figure 2-3.

The NBS architecture proposes a layered approach to systems design, which is not unlike the ISO/OSI layered approach adopted by the MAP/TOP standardization efforts. The advantages of the layered approach are clear. It allows the manufacturing system to be modularized and segmented. At the same time, each module forms part of the whole manufacturing system jigsaw. In general, it is a widely accepted architecture and, like ICAM, some of its concepts can be used by manufacturing system designers. However, like ICAM it is also nonspecific for a particular type of manufacturing system.

## CAM-I Architecture

The CAM-I architecture was developed by CAM-I Inc. (Computer Aided Manufacturing - International), a nonprofit organization in the United States founded in 1972 to promote collaborative research among companies with common interests in CAM.[35, 36]

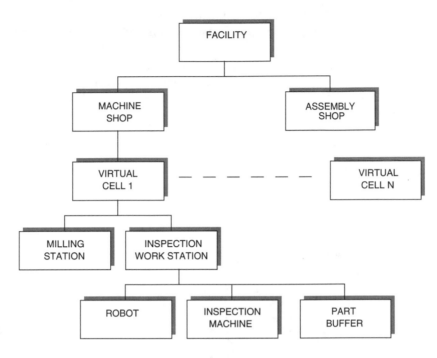

**FIGURE 2-3.** NBS control architecture.

CAM-I, a consortium of industries and research institutes, has developed the CAM-I CIM enterprise model, which again generically represents the manufacturing enterprise, and through breakdown into functional areas aims to tackle various issues surrounding CIM. The areas that have been covered in this initiative are (1) company policies and procedures, (2) organizational structure, and (3) standards beyond STEP/PDES. Similar to CIM-OSA, CAM-I views the enterprise from five different perspectives: (1) management, (2) information, (3) function/activity, (4) computer systems, and (5) physical structures. CAM-I's CIM model is continuously evolving and offers companies not only a reference model but also potential standards and policies.

### CIM-OSA Architecture

CIM-OSA or CIM-Open Systems Architecture from the ESPRIT-funded consortium AMICE (a group of major European companies and research institutes) aims to develop *an all-embracing conceptual framework* within which companies can implement CIM.[37, 38] CIM-OSA is a reference model, meaning that it will not emerge as a standard but that standards can evolve from it, in much the same way as MAP evolved through ISO's OSI architecture. CIM-OSA is a complete description of a manufacturing enterprise using various representations such as Organization, Resource, Information, and Function (see Fig. 2-4). It describes, using these representations, each function in the enterprise in generic form. The areas within the scope of CIM-OSA are (1) Product

Information, (2) Manufacturing Planning and Control Information, (3) Shop Floor Information, and (4) Basic Information which includes standards and guidelines for the company.

A very important aspect in the CIM-OSA project, which to some extent mimics the IDEF approach by ICAM, is the development of a comprehensive methodology for modeling and specifying the CIM architectures being developed within the project. This methodology is currently a CEN/CENELEC prenormative standard. The expectation of CIM-OSA is that in the future it will become a full standard for the way in which manufacturing systems are modeled.

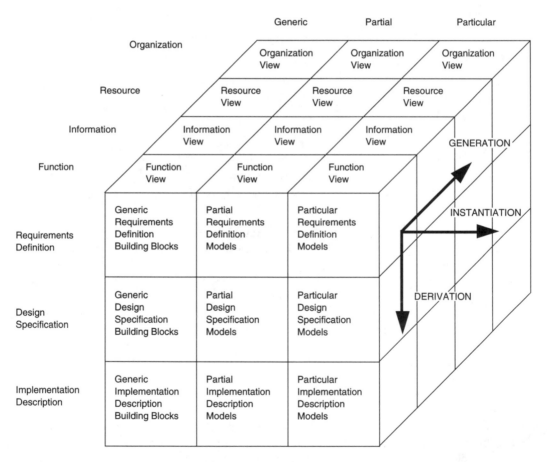

**FIGURE 2-4.**   CIM-OSA modeling cube.

The cube is a diagrammatic representation of the various modeling views identified in the project. In one view, modeling is described as consisting of generic, partial, and particular requirements definition stages. In another view, modeling has function, information, resource, and organization levels of analysis. Finally, in a third view, modeling is described as having requirements definition, design, and implementation stages in development. The combination of each of these three views gives 36 different combinations of the modeling domain. In theory, each of the 36 can utilize 36 different modeling tools or methodologies, but the consortium has reduced this to a small number of common modeling methodologies.

## IMPACS Architecture

The IMPACS architecture has been developed by a European consortium of industries and research institutes. The architecture outlines a hierarchy of business functions from Business Planning through Master Production Scheduling and on to Shop Floor Control.[39] The hierarchy is illustrated in Figure 2-5 and shows a number of generic functions that act as decision support for production management. A key feature of the architecture is the allocation of tasks at the lower levels in the hierarchy. At this level, operational issues are addressed through what is called Production Activity Control (PAC).

In essence, the factory floor may be seen in terms of a number of production cells. Each cell is modeled on group-technology principles and contains the machines and handling equipment necessary to carry out specific production processes. The cells are controlled by PAC, which consists of an individual scheduler, dispatcher, mover, producer, and monitor. Coordination of materials and schedules between individual PAC cells is done through a Factory Coordinator or super-PAC system.

The Factory Coordinator carries out essentially two functions. The first is coordination between individual PAC cells - on a principle similar to PAC itself (i.e., Factory Level Scheduler, Factory Level Dispatcher, and Factory Level Monitor). The second is a production environment design function. The IMPACS architecture recognizes the need for production environments to be constantly redesigned in order to meet the demands of market and resource constraints. The basic concept behind the IMPACS architecture is that each of the functional blocks (i.e., Dispatcher, Factory Level Scheduler, etc.) needs to be designed as autonomous yet integrated software modules. In the marketplace, suppliers of factory software may become expert in the development of any one in particular. Yet, because of the nature of the way each module is architecturally designed, modules from different suppliers can easily be plugged into the overall IMPACS structure.

The development approach used by the consortium in creating IMPACS incorporates five development tools, each selected for their part in a reference model for systems design. This reference model identifies the need to model function, information flows, decisions, data, and physical entities. The individual methods selected are IDEFo, Data Flow Diagrams, GRAI, IDEF1x, and Group Technology (details on each of these methods are given in Section 2.3.1). The IMPACS architecture is widely accepted among software vendors and manufacturing industry as a useful and practical interpretation of the production management system. Not unlike many of the other architectures discussed, IMPACS popularity and future will depend on systems suppliers adopting its principles and supplying the market with compatible products.

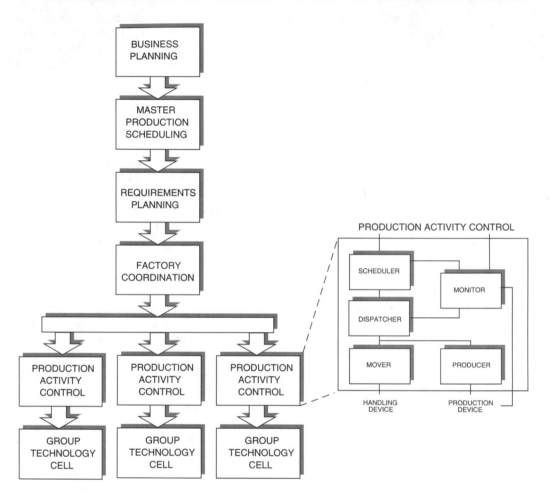

**FIGURE 2-5.**   IMPACS model.

## CALS

CALS (Computer Aided Logistic Support) was a project developed at the U. S. Department of Defense. Its purpose, like many of the other architectures, was to describe a generic factory in terms of functions and information flows. The major emphasis in the CALS project was on standards development. Many new standards have already evolved from CALS development. By *mandating* that contractors use these standards, the U.S. Department of Defense aims to achieve more efficient and automated operations. The areas within the scope of CALS include: (1) CAD data exchange using the PDES standard, (2) automated publishing, and (3) file management. The CALS initiative is not only limited to contractors for military contracts. The results are applicable across a wide range of industries. More recently the initiatives created through CALS have been extended to cover projects relating to the development of wide-area networks specifically designed for allowing vendors, suppliers, dealers, and customers to access each other via electronic data interchange.

## 2.2.3 Computer Manufacturer Models

The second category of architecture and model type is that created by computer manufacturers. Because computer manufacturers still by and large dictate the available standards and products on the information technology market, they are a useful source of information for the manufacturing system designer. In literature searches, two interesting models from the two leading computer manufacturers were found. Both of these models would seem to have benefited from the advantage of being developed by small focused groups within each of the organizations in question. Each model is very clear and precise. But they also suffer as a result of this small focus. Not surprisingly, each model reflects well-known product development concepts within the organization in question, and in the case of one, the model borders on being a marketing brochure. Both of these models will be discussed later.

A third architecture from Hewlett Packard (HP) is called OPENCIM which HP is releasing under the banner of "New Wave Computing Architecture."[40] On examination of currently available descriptions of the architecture, it would appear to be HP's response to the MAP/TOP standardization issues discussed earlier. Rather than being an architecture for manufacturing or CIM, it focuses primarily on data integration issues. The architecture also touches on information integration issues in the upper layers of MAP and TOP. HP for example will implement the MMS protocol into the architecture but what is not clear is what hardware or shop-floor equipment this protocol will service.

### IBM Model

The IBM model is presented in an elaborate publication which attempts to provide its reader with a view of CIM through the use of sophisticated block diagrams and icon-based computer architectures.[41] The entire text is very lavishly annotated with color diagrams, thought-provoking quotes, and tidbits of useful information about CIM. The heart of the publication is the so-called "architecture of CIM" which is described as consisting of six primary functions. These functions and associated subfunctions are illustrated in Figure 2-6.

An interesting aspect about the IBM model is that although generic, it offers an excellent guideline to the creation of computer (hardware) architectures in manufacturing.

### DEC Model

The DEC model entitled "Reference Model for Computer Integrated Manufacturing," is quoted by the authors as being "a framework for supporting design of CIM systems in a modular fashion across Digital."[42] It is a comprehensive reference model which illustrates all of the functions of manufacturing and the so-called supporting technologies. All of these manufacturing functions are grouped into three interrelated areas—Design to Build, Production Systems, and Business Systems—as illustrated in Figure 2-7.

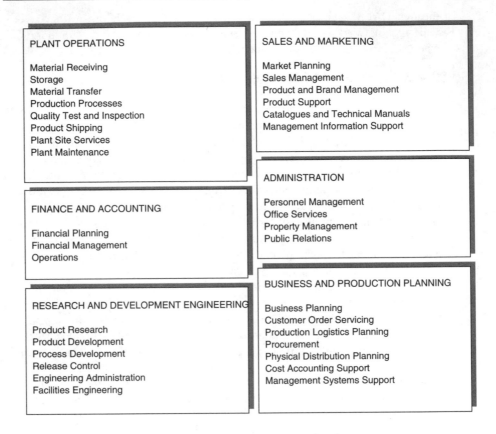

**FIGURE 2-6.**   IBM CIM architecture functions.

Each of these areas are detailed with respect to their activities and the relationships among activities. The models are created using conventional block diagramming techniques. Although carefully detailed, the block diagrams do not indicate major information flows. To learn about information flows, the reader must sift through the accompanying text. An interesting aspect of the model is the development of a multilayer approach. Three layers (illustrated in Fig. 2-7) are generated. As explained by the authors "the concept of layering is applied to partition system components for ease of implementation and to provide transparency of underlying technology from the higher layers."

In essence, the top layer is roughly based on the functional aspects of the manufacturing system described. The two lower layers from Digital's perspective represent the computer hardware/software (support layer) and communications hardware/software (service layer) that allow the system to be implemented. It is therefore claimed by the authors that all three layers can be worked on in isolation during the architectural design stages and knitted together later during implementation. For the manufacturing system designer, these approaches to multilayering are very interesting. The approach promotes the concentration of functional issues by experts in the functional area and leaving details of the computer systems to other personnel who are perhaps more qualified.

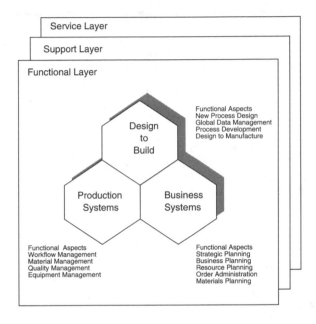

**FIGURE 2-7.** DEC model–main functions.

## 2.2.4 Individual Models

The final category of architecture and models are those developed by individual researchers. Clearly there are many potential architectures which can be described here. However, two—Harrington and Scheer (mentioned earlier)—have been selected for their particular insights into this whole area.

Thacker and Ranky have also presented their own versions of the CIM architecture or model. Thacker developed his model "CIM Model Architecture" in conjunction with the Society of Manufacturing Engineers (SME) in the United States.[43] In the model, Thacker divides the activities of manufacturing along two axes. The first is along functional or perceived functional boundaries such as design, production, sales, and so on. The second is along management and control axes where each function can be described as comprising general, functional, operational, and production activities. The result is called the CIM cone which is claimed in one source by Thacker as being "a practical guide for planning, designing and implementing CIM."

Ranky developed his CIM model in parallel with his development of a number of suites of software.[44] This software, he claims, offers CIM solutions to a wide domain of manufacturing applications. The model is based on conventional block diagramming techniques. This technique of modeling is unfortunately ambiguous and needs to be supported with considerable text. The supporting text describes each function and its main interactions with other functions as well as with the various databases conveniently defined for storing manufacturing information. This model has the advantage over other models in being supported with examples of software systems and implementation guidelines for the CIM technologies such as robotics, CNC, and coordinate measuring machines.

A-11 Manage Enterprise              A3 Produce Products
A-12 Manufacture Products           A31 Plan for Manufacture
A-13 Market Products                A32 Make & Admin. Sch. & Budgets
A-14 Support Corporate Activities   A33 Plan for Production
                                    A34 Provide Production Resources
A0 Manufacture Product              A35 Obtain Production Materials
  A1 Manage Manufacturing           A36 Convert Materials to Products
    A11 Plan Projects                 A361 Control Production
    A12 Make Project Schedules        A362 Make Parts
    A13 Administer Projects           A363 Make Sub-assemblies
  A2 Develop Projects                 A364 Make Final Assemblies
    A21 Develop Conceptual Design     A365 Test Products
    A22 Develop Preliminary Design    A366 Ship Products
    A23 Develop Detailed Design     A4 Support Service of Products

**FIGURE 2-8.**   Node diagram for Harrington's architecture.

## Harrington Model

Harrington developed his model of the integrated manufacturing environment using the structured analysis tool IDEFo.[18] The model looks in detail at 24 activities of manufacturing which Harrington was interested in discussing. These activities are listed in the IDEFo node diagram in Figure 2-8. In his model Harrington easily identifies three separate flows of information for his reader. The first flow is associated with the functional structure of manufacturing indicating the control flows on each of the activities and the major data flows among activities. The second flow is the routine feedback flow which illustrates the control information both generated and used by the various activities. Finally, he identifies a third flow which he calls the feedback of problems. In identifying this flow, Harrington acknowledges the need for more integration between upstream and downstream activities.

An interesting aspect about Harrington's model is that it is surprisingly easy to follow and understand, and the information flows are easy to trace and verify. This can mainly be attributed to the IDEFo modeling technique which Harrington uses for modeling. Another reason, of course, must be attributed to Harrington's dedication for making what he calls the "art" of manufacturing into a science. Harrington is not shy about stating how useful he believes his model is: "the structure set for this model is common to all manufacturing enterprises; furthermore, I believe it is reasonably invariant with respect to time, so that we may rely on it for guidance in the future."

Harrington's model is of course generic and may be too general for most systems design projects. However, the approach and ability to manipulate the IDEFo methodology to describe what is a complex system provides designers with a useful lesson in systems analysis, architecture, and model development. The model, with its classification of information flows, is also useful in allowing designers to appreciate the holistic nature of manufacturing systems. Figure 2-9 illustrates part of Harrington's model, created for describing the interaction between management, product development, production, and product-support activities in manufacturing.

Each of the four activities in Figure 2-9 represent at a general level all of the activities required to carry out the manufacture (design and production) of products. Through breakdown of each of the activities (boxes) illustrated more detailed activities will be illustrated which act in concert to produce the product concepts, products, and parts.

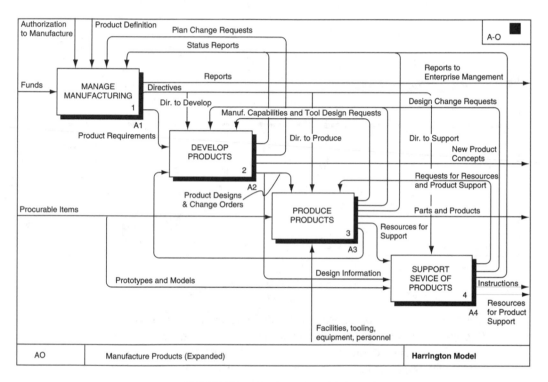

**FIGURE 2-9.** Harrington's model shown in part.

## Scheer Model

Scheer is a German engineer whose research work on CIM has been published in a series of well-received books. In one of his books, *CIM - Computer Steered Industry*—he presents a model of CIM.[19] In it he identifies the information in manufacturing as consisting of essentially two flows. The first is the order information flow as in production planning and control (PPC), and the second is the technical information flow as in CAD/CAM. One way of illustrating these flows is shown in Figure 2-10.

Scheer develops his entire argument for manufacturing systems integration around this reference model that makes extensive use of the concept of distributed databases which is the first of two reasons why this model is different from the others mentioned.

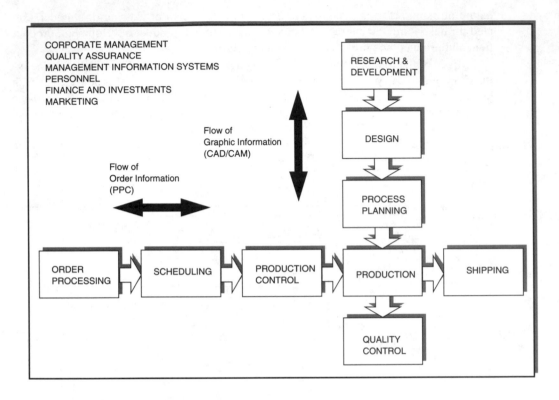

**FIGURE 2-10.**   Graphic and Order Information Flows.

The topics of data structures, storage, and management are given a detailed analysis in this model. While at times this analysis becomes difficult to read, it is an excellent model by which to compare alternative manufacturing architectures. The second reason the model is particularly interesting is Scheer's concept of dividing information into two interrelated streams—PPC and CAD/CAM. Defining the model around this separation allows the reader to follow a logical sequence of information flows. In this way, for example, the reader can see exactly what may happen to an order or to a new product in the manufacturing system. This approach, while not as structured as the IDEFo approach by Harrington, can be very effective. Another German source found to use this technique described the flows "Geometry Oriented Data Flow" and "Administration Oriented Data Flow."[45] The discussion in this reference also relies heavily on the two flows for isolating issues in manufacturing systems.

## 2.2.5 Architectures and Manufacturing Systems Design

Each of the above architectures or models offer a number of opportunities for the manufacturing systems designer. On the one hand, they allow designers to have an insight into how manufacturing systems elsewhere have been designed to operate or in most cases how they should ideally operate. They also offer the designer an insight into the practical aspects of the architectural development approach. Each of the architectures discussed above has used such an approach in one form or another. Many of the projects (ICAM and Harrington, in particular) have complemented this approach with the use of languages or tools using the concepts of structured analysis. This is clearly another lesson from this whole area for designers. In both the ICAM and Harrington approaches, IDEFo was chosen as one of the main communication mediums.

All of the architectures discussed are to some extent idealistic. In all cases, they do not represent real manufacturing systems. In this respect they more than likely will not assist systems designers with the operational details of their own system. The main advantage of reading these architectures must be the insight gained from the approach and perspective adopted by their authors. Since these perspectives are so numerous, a generic approach to modeling the perfect manufacturing factory is almost impossible. Instead what many companies and researchers are turning their efforts to is using tools and methods that facilitate the development of their own system architectures and models. These tools allow the designer to create his or her own so-called factory of the future which takes into account the real constraints of their own manufacturing environment.

## 2.3 SYSTEM MODELING TOOLS AND METHODS

In the section on System Architectures and Models (Section 2.2), reference was made to tools such as IDEFo and those used in CIM-OSA which facilitate the conceptual modeling approach to manufacturing systems. These tools are referred to here as System Modeling Tools and Methods. For clarity, System Modeling Tools refer to techniques used for diagrammatically representing functions or activities. As is the case with most of these tools, there is an associated methodology for their implementation. As this methodology is often more important than the tool, it will be referred to in this section as a System Modeling Method. As the reader will notice, some of these methods have no unique system modeling tools of their own, but rather use other well-proven system modeling tools. ICAM, for example, developed the IDEF modeling method in which it used three well-proven system modeling tools called IDEFo, IDEF1x, and IDEF2. In effect, the IDEF method was a shell into which three tools could be placed.

There are a large number of so-called system modeling tools and methods available. Many of these tools facilitate a structured approach to the modeling of manufacturing systems. These tools are by no means new. Some of the tools have been in use in software engineering for a number of decades, and indeed the area of CASE or Computer Aided Software Engineering is very similar to the area being discussed in this section. However, here the emphasis will be on tools and methods for manufacturing systems design and not exclusively software systems design.

System Modeling Tools and Methods allow designers to share a common perspective on a proposed system's functionality. The advantages of using these tools and methods are clear. Along with promoting a common perspective on manufacturing systems, they also allow unambiguous definition of various models and architectures that contribute to the development of the integrated manufacturing environment. A large number of structured methodologies now exist for the development of manufacturing systems. Figure 2-11 shows a general classification of these tools.

Tools and methods in the *Decision* category are those tools that are used for modeling the decision-flow process. Here two tools are to be found, both of which have been developed at the University of Bordeaux by the GRAI laboratory. These tools are widely used within many of the ESPRIT projects referred to earlier and are dealt with in a later discussion. The *Activity* category has tools and methods that model the activities of a manufacturing system. The most important tool here in the context of this book is IDEFo. Other tools in this category include SADT or Structured Analysis and Design Technique which is the technique from which IDEFo was developed. JMA from James Martin Associates is an activity modeling tool used primarily by the consulting firm of the same name. The other three activity modeling tools—CORE, A-Graph, and INFOREM—are similar to IDEFo, although they have not gained the same popularity.[46] The reasons for this are not clear. One reason may be that although comprehensive, IDEFo is essentially a simple tool to adopt and use. CORE, for example, proposed by Mullery and developed by British Aerospace is described as a much more comprehensive and detailed methodology than IDEFo.[47, 48]

| CIM DESIGN TOOLS AND METHODS | | | | | |
|---|---|---|---|---|---|
| DECISION | ACTIVITY | DATA | DATA FLOW | PROCESS LOGIC | USER INTERFACE |
| GRAI NET<br>GRAI GRID | IDEFO<br>SADT<br>JMA<br>CORE<br>A-Graph<br>INFOREM | IEM<br>IDEF1x<br>NIAM<br>EXPRESS-G<br>EXPRESS<br>ACM/PCM<br>DADES | CIAM<br>ISAC<br>JSD<br>SDM<br>SASD<br>SSA | Structured<br>English<br>Warnier<br>Orr<br>Petri nets<br>IDEF2<br>SIMAN<br>SLAM | USE<br>ISSM<br>VUIT<br>OSF-Motif |

$$\longleftarrow \text{———— GIM ————} \longrightarrow$$
$$\longleftarrow \text{——— IDEF ———} \longrightarrow$$
$$\longleftarrow \text{—SSADM—} \longrightarrow$$
$$\longleftarrow \text{—MERISE—} \longrightarrow$$

**FIGURE 2-11**   General classification of System Modeling Tools and Methods.

The next two categories in the classification are, *Data* which represents those tools that are used for modeling data structures, and *Data Flow* for tools that model data flow among activities or functions. There are a large number of tools in this area. Only a select number of the more commonly used tools will be discussed in a later section. The tools SSA, EXPRESS, and SSADM have gained in popularity for use in the development of manufacturing systems. The two final categories of *Process Logic* and *User Interface* have been included here because they complement the manufacturing systems design process. A more comprehensive analysis of System Modeling Tools and Methods, particularly in the area of data flow and data structure, is available from Rock-Evans in report format.[46]

## 2.3.1 System Modeling Tools

Three tools, one from each of the main categories above, will be described generally in this section. These are GRAI, SSA, and EXPRESS. Particular attention was given to the description of IDEFo in Section 1.6, as this tool will be used later in the development of the design road map.

### SSA

The Structured Systems Analysis (SSA) tool was originally developed by Gane and Sarson for use in modeling data flows.[49] These researchers have also been equally effective in representing the flow of physical entities. SSA is based on similar principles to IDEFo in that it modularizes functions and uses the principles of hierarchical composition. However, it also differs in many respects from IDEFo. It is more detailed, using seven different types of box as well as symbols for data stores, files, and buffers. An example of a data-flow diagram is given in Figure 2-12.

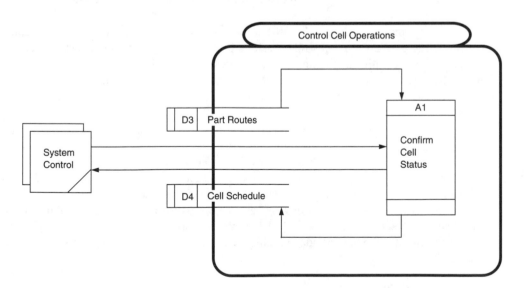

**FIGURE 2-12.**   Example of a data-flow diagram.

The methodology used in the development of the diagram can be summarized as follows:

1. Identify the external entities.
2. Identify the main activities performed within the systems boundary.
3. Identify the inputs and outputs.
4. Draw activities and diagrams.
5. Refine models based on additional knowledge.
6. Produce lower-level explosions.

In general, the method is very widely accepted and has been utilized as part of many of the methodologies given in the earlier classification. It can be said to be more software oriented than IDEFo as many of the symbols represent physical software items. In this respect, it has been developed into a so-called CASE tool whose output is source code in a particular programming language. In many of the articles reviewed, SSA is the preferred modeling tool over all others available.[50, 51, 52] In some of these references the IDEF trilogy has been emulated, although IDEF1x has been replaced by SSA or an SSA clone.

### Grai Grids and Grai Nets

The GRAI method, developed at the GRAI laboratory in France, has been developed to model decision-flow processes in manufacturing environments.[53, 54] While it can be applied across the manufacturing system, its particular application domain is in the modeling of production management systems. The GRAI method uses two tools called the GRAI Grid and the GRAI Net. In the GRAI Grid various activities are modeled with respect to the decision and information flows between them. The GRAI Net, on the other hand, examines each of these flows in detail and through the use of graphical modeling explicitly defines relationships between inputs, outputs, and the decision process. These relationships can be later expressed through mathematical representations, thus making it easier for the designer to explicitly define information and decision processes. An example of the GRAI Grid and the resulting GRAI Nets is given in Figure 2-13.

The GRAI technique (grids and nets) gets much of its theoretical background from work in systems analysis by Boulding and work in hierarchical systems by Mesarovic, Macko, and Takahara.[11, 55] The use of a GRAI conceptual model helps guide the systems designer in understanding the context for manufacturing systems to which the GRAI technique is applied. In the conceptual model four systems are identified: (1) the Decision system, (2) the Operating system, (3) the Information system, and (4) the Physical system. Researchers at GRAI believe that understanding these four types of systems facilitates the better understanding and modeling of manufacturing systems as a whole.

### Express/Express-G

EXPRESS is a tool used to model data structures and information flows in information systems. EXPRESS is an ISO standard (ISO 10303) for global manufacturing

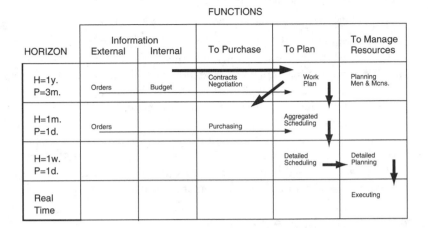

(a)

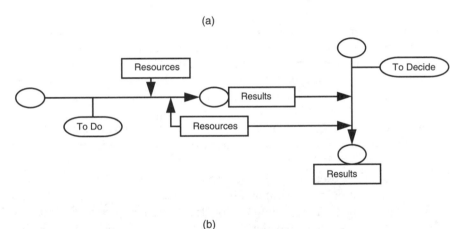

(b)

**FIGURE 2-13.**   Example of (a) GRAI Grids and (b) GRAI Nets.

programming languages developed in response to a need for an explicit data-modeling tool used in the development of projects such as STEP and IGES/PDES. The EX-PRESS language has two parts. The first part is the EXPRESS language which is used to model information. The second part, EXPRESS-G, is a graphical subset of the EX-PRESS language promoted as an easy-to-use graphical tool for the implementation of EXPRESS. The EXPRESS language is more explicit than the modeling tools discussed thus far, reflecting the relative position of EXPRESS as lower in a modeling tools hier-archy. The lower in the hierarchy one goes, the more explicit the modeling language becomes, until theoretically one reaches actual software code or work procedures. The design goals behind the development of EXPRESS are:

1.  A modeling language parsable by computers as well as humans

2.  A language designed to partition diverse systems and functions

**3.** A language that identifies entities of things

**4.** A language that avoids specific implementation views

An example of an EXPRESS-G model and its associated code is given in Figure 2-14. In EXPRESS entities are defined in terms of attributes, the traits or characteristics considered important for use and understanding. These attributes have a representation which can be a simple data type (integer) or another entity type. The EXPRESS language has been developed around principles used in well-known languages such as ADA, C++, Pascal, and SQL among others. Within many collaborative EC-sponsored research projects, EXPRESS is gaining in popularity, particularly since its adoption as an ISO standard. In the CIM-PLATO project to be discussed later, for example, EXPRESS has been used in conjunction with IDEFo and OSF-Motif for the definition and integration of software among a number of software developers.

## 2.3.2 System Modeling Methods

In this section system modeling methods, which combine a number of system modeling tools, will be described. In the previous section a number of system modeling tools were introduced as popular representatives of a large number of tools available. Each of these tools has an approach or structured technique by which it can be implemented. For example, IDEFo uses what Ross calls a "reader-author cycle" approach for modeling systems. The definition of tool and method is only to help distinguish between stand-alone tools and combinations of tools, respectively. In many of the methods discussed, the tools used will not be outlined, and for more information the reader is referred to various references.

As outlined earlier, IDEF was developed in the ICAM program in order to help designers answer three questions about manufacturing systems: (1) What activities are being performed? (2) What information and data are needed to support these functions? (3) What changes in the functions and information occur over a period of time? In order to answer these three questions, IDEF developed the three languages IDEFo, IDEF1x, and IDEF2. Each was selected to model activities, data structure, and state transitions, respectively. The IDEF methodology saw all three modeling tools being used in series and all being linked so that information could be shared. Only IDEFo would appear to have stood the test of time. IDEF1x is regarded as a useful tool in data structure design in the creation of databases. However, it is criticized as not being useful for data-flow modeling.[52] IDEF2 is regarded by a number of sources as being of limited use, difficult, and time consuming. Because of the large number of other tools available IDEF1x and IDEF2 have not found significant application outside of the IDEF projects.

The Structured Systems Analysis and Design Methodology (SSADM) is a development by the U.K. government body Central Computer and Telecommunications Agency (CCTA).[56] It has been developed into a standard for use in British Government software projects. The methodology uses three types of techniques. One based on the SSA modeling tool describes data flows. The second describes data structure (in much the same way as IDEF1x or Express), and the third technique describes entity life histories. This methodology is clearly important in that software houses must use it to secure government contracts in the United Kingdom. Manufacturing systems designers may well learn to adopt this approach and insist that their software suppliers and systems integrators also adopt the methodology.

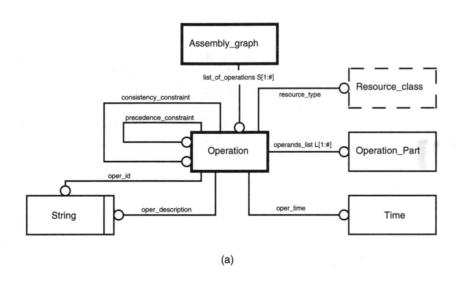

(a)

```
SCHEMA assembly_graph

TYPE frame = array [1:4, 1:4] of REAL
END_TYPE

ENTITY assembly_graph
 list_of_operations  SET [1:#] OF operation
END_ENTITY

ENTITY operation
 oper_id
 oper_time
 operands_list     SET [1:#] of operation_part
END_ENTITY



    .
    .
    .

END SCHEMA
```

(b)

**FIGURE 2-14.**   Example of (a) EXPRESS-G and (b) EXPRESS languages.

In the ESPRIT CIM-PLATO project to be discussed later in this chapter, a loose methodology has been adopted which requires members of the project group to use the three following system-modeling tools: IDEFo, EXPRESS, and OSF-MOTIF. This approach, they argue, will make it easier for true information integration of the various software modules being developed. A similar approach is being used by the ESPRIT IMPACS project where the tools GRAI, IDEFo, IDEF1x, and Group Technology are being used as part of a systems-modeling method. This combination is called the GRAI Integrated Method (GIM) and is currently being refined by the GRAI laboratory. ESPRIT CIM-OSA, as mentioned previously, has also developed a modeling method based on a modeling framework called the CIM-OSA cube. The use of the cube results from the use of various system-modeling tools.

Finally, there are a number of other methods available which vary considerably with regard to their approach and the tools they use. In Germany a methodology is being developed which includes a number of the tools mentioned above, as well as some new ones in what the developer calls an "objected oriented approach to systems design." The method called MOOD is being used in CIM-OSA as well as in a number of German projects.[57] The "M" approach developed in Italy considers organizational as well as structural issues in manufacturing. Some of the tools used include petri-nets, but there is also a larger emphasis on organizational modeling issues.[58]

## 2.4   SYSTEM PLANNING TOOLS

In the previous section, tools and methods for modeling manufacturing systems were described. These tools were characterized by the various diagramming techniques that each employed for creating conceptual models. In this section, so-called system planning tools will be described. These tools can be described as those used in planning systems development activities, using a mixture of mainly technical and some social ideas. They focus more on the activities of systems design than on the modeling process. Three tools are identified that can be conveniently described as tools or methodologies for manufacturing systems designers. These tools (CIMPLAN, ISP, and Functional Analysis) are methodologies that border on being techniques in that they have some tools for system modeling. In fact, one of the tools, ISP, links into the CASE-type tools mentioned earlier.

### 2.4.1 CIMPLAN

One of the first areas for the effective use of tools and methodologies within the context of manufacturing systems design is that of planning of manufacturing projects and their step-by-step implementation into the manufacturing environment. CIMPLAN is a comprehensive and systematic methodology that facilitates the process of strategic project development and implementation.[25] CIMPLAN is developed to fit within the framework created by the strategic planning hierarchy of Corporate Strategy, Business Strategy, Manufacturing Strategy, and CIM Strategy. It develops its primary guidelines around the critical success factors important to a particular company. These factors

(Capacity, Quality, Response Times, etc.) determine how and when projects are to be implemented. The methodology is based on a systematic six-stage process, with each of the stages concisely described and augmented with various charts and diagrams. These six stages are as follows:

***Set Meaningful Objectives:*** Here the user of the tool is encouraged to develop objectives that reflect the overall goals of the company. The user is shown how objectives such as "Improve Product Quality" can be derived from the company's strategy, and be achieved through a number of different projects.

***Audit Current Conditions:*** In this stage the user is shown how to audit the condition of the existing system in general terms. Prioritization of areas for development within the factory is achieved through careful analyses of the audit.

***Define Specific Projects:*** In this stage specific projects are identified as a result of the goals set and priorities given to areas within the plant from the previous two stages. The definition of the projects takes place using structured methodologies to allow better *understanding* of the project.

***Follow a Logical Sequence:*** Here the user is shown how to plan various projects over a typical time horizon. The technique used is general and encourages the user to be initially rough and general at this level of planning. At a later stage the fine detail can be added. The author uses the practical advice *crawl, walk, run* to carry home this concept.

***Assign the Proper Team:*** This stage corresponds to the initial phases of project implementation. Advice is given on project team structure and possible pitfalls.

***Monitor Plans and Results:*** The final stage is to monitor the plans made for progress and the results being achieved during project implementation. It outlines the difficulties often experienced by projects during the so-called "Dip," where the summation of problems can sometimes cause project failure.

The CIMPLAN methodology, although general, is extremely easy to understand and to implement. What it lacks in detail, it makes up for in short snappy words of practical advice from the author. Rather than take the approach of telling project managers what to do, it prefers to offer guidelines and useful facts and information about manufacturing systems design in general.

## 2.4.2 Information Strategy Planning

Information Strategy Planning (ISP) is a methodology developed by James Martin Associates.[59] It is part of JMA's Information Engineering Methodology which constitutes a suite of tools and methods for information systems analyses. The complete ranges of tools and techniques are defined through the deliverables that they provide. These are illustrated in Figure 2-15.

The ISP element can be seen as a project development stage of the detailed information analysis stages. In the brochures, ISP is described as providing "details of the underlying infrastructure and plan of activities to meet the information system needs."

ISP is designed to provide what JMA calls the Information, Systems, and Technical architectures for the system being analyzed. It does this after it has initially defined the strategies and policies of the company. This is followed by an outline of the business requirements of the company. The ISP approach seems at first glance to be the ideal solution for a methodology to aid manufacturing systems designers. But on more careful examination, it can be seen to be very much a software engineering approach. It describes information in terms of well-defined data terms according to the JMA software engineering procedures. In reality, of course, manufacturing systems are more complex. However JMA's ISP approach is relevant to the issue of information integration.

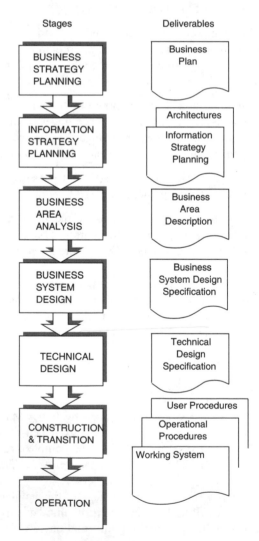

**FIGURE 2-15.**   Deliverables from JMA's Information Engineering Methodology.

## 2.4.3 Functional Analysis

In his book *Functional Analysis—Simplify Before Automating,* Eugene Wittry outlines the technique of functional analysis (FA) for planning systems in organizations.[60] Wittry developed FA from his work on Management Information Systems at Bradley University and according to him it is a technique "based on the idea that users are responsible for designing their own systems." These same systems need not necessarily require automation or computerization. Wittry argues that FA is more applicable for simplifying and integrating information needs than for automating them.

FA consists of 14 almost sequential and sometimes parallel steps. Each of the steps allows designers to analyze systems requirements, beginning with a description of the business and moving sequentially through areas such as identifying business strategies, mapping information flows, and finally determining possible computer systems requirements. Each of the 14 steps are given the following titles:

1. Business Description
2. Business Mission
3. Measures of Business Performance
4. Business Goals
5. Business Strategies
6. Department Mission
7. Measures of Department Performance
8. Department Subfunctions
9. Department Strategies
10. Organizational Conflict
11. Information Flow Chart
12. Subfunction Description
13. Rules to Resolve Conflicts
14. Computer Systems Requirements

The list is comprehensive and almost seamless. FA approaches the development of systems in a very similar way to JMA's ISP. On close examination it is a mix of sound social and technical planning concepts, all built on the need for change to emanate from a business strategy. While it is shallow on theory and the use of modeling tools and methods, it presents a holistic and structured approach for system analysis.

## 2.5   STANDARDS

Standards, both existing and emerging, are an important source of technical planning information for the manufacturing systems designer. The use of standards ultimately reduces the overall cost for the development of a new system. Standards offer designers guidelines for the functional operation of manufacturing systems. Once a system is said to adhere to a specific standard, the purchaser of that system has a clear idea of what he or she is getting. Standards also break the dependence on buying hardware and software from a single vendor. They make it easier for system designers to choose subsystems from a number of vendors. In recent years, perhaps the most widely known emerging standard has been MAP (Manufacturing Automation Protocol), which is one of the new standards for the definition of communication protocols among dissimilar computer systems. While MAP has received much attention in the recent past, it is only one of a very large number of standards to which the systems designer has public access. While some of these standards can be said to be important aids to systems design, others are essential to the creation of the integrated environment. As mentioned earlier, this integrated environment is now a key to success for many manufacturing businesses.

Many of the more important standards to emerge in the last ten years have stemmed from the fact that most, if not all, manufacturing functions manipulate stored information. The computer, which is the common element of most manufacturing equipment, can be said to form the building blocks upon which the modern manufacturing enterprise is built. In the area of information exchange and data handling three categories of standards can be identified: (1) Communications, (2) Graphic Interface, and (3) Hardware. For the system designer other standards also exist that can enable the specification of systems. These can be classified into four additional categories: (4) Reliability, (5) Safety, (6) Quality, and (7) Performance Measurement. Each of these categories is illustrated in Figure 2-16 with a brief description of the areas it covers.

| INDUSTRIAL STANDARDS | |
|---|---|
| CATEGORY | DESCRIPTION |
| COMMUNICATIONS | Standards which cover the transmission of data. e.g. Protocols. |
| GRAPHIC INTERFACE | Standards for the exchange of graphical information. e.g. Graphic Processors. |
| HARDWARE | Standards for hardware. e.g. Computers and CNC Machines. |
| RELIABILITY | Standards which cover the reliability of hardware equipment. |
| SAFETY | Standards which cover safety in the manufacturing environment. |
| QUALITY | Quality standards for hardware and software products |
| PERFORMANCE MEASUREMENT | Standards for performance measurement of Hardware, Software and Systems. |

**FIGURE 2-16.**   Standards categories for manufacturing systems design.

## 2.5.1 Communications Standards

Standards in this area can be said to have been dominated by the developments of MAP and TOP (Technical Office Protocol) through the International Standards Organization's OSI (Open Systems Interconnect) architecture. These standards initiatives, which started in the early 1980s, have stimulated worldwide interest. Although still evolving, MAP and TOP offer an exciting prospect for addressing some of the Data Integration issues mentioned earlier. The major objective of MAP and TOP is the development of specifications for communication protocols among dissimilar computers. The access and adherence to these specifications will allow Open Systems Interconnection on the shop floor (MAP) and in the office (TOP). Many of the designations for standards adhering to the OSI/ISO architecture in the MAP and TOP domain are given in Figure 2-17.

The latest versions have allowed the development of a wide variety of products that successfully implement the functionality of a number of the layers in the OSI model. With these products the manufacturing system designer has at least the option to choose the MAP or the TOP computer network. The designer also has the option to choose different computers that plug into the network via special cards. Then by requesting from his or her application software vendor that they develop the software to link into the upper layer of either MAP or TOP an open system is begun. It is this link that defines the boundary between what was earlier described as data integration and information integration. As suggested above, MAP/TOP are the only real OSI alternatives available. Information integration for the system designer is an issue of precisely what information to send through the communication network. It is a much larger issue since the definition of the information must on the one hand satisfy the planning and control needs of the manufacturing system, and on the other hand it must allow OSI between different application programs.

A significant MAP/TOP community development is the Manufacturing Message Specification (MMS) which normally resides as a MAP layer 7 entity and is depicted in Figure 2-17.[61] MMS has evolved to offer kernal message handling services.[62] It defines message shells where there are two specific areas or blocks. The first area allows for a command such as "download data," and the second area allows for the data being handled. By defining a set of application specific commands, application programs can be developed or modified to interface with the MMS layer in the MAP network. The significance of this interfacing will become clearer as we move more toward the discussion of the modeling of information flows through the use of System Modeling Tools and Methods.

Although MMS is applicable to a wide range of applications, it does not define all command and message types. The intention is that different companion standards be developed for specific commands and message formats. Each companion standard should cover the specific commands and message format for a specific shop-floor application. Companion standards would therefore be written by those industries and standards organizations most familiar with their application. For example, a companion standard is currently being developed under the DIN association for the commands and message format of Robot controller communications. This is through the ESPRIT project CIM-PLATO which is discussed later in Section 2.6.

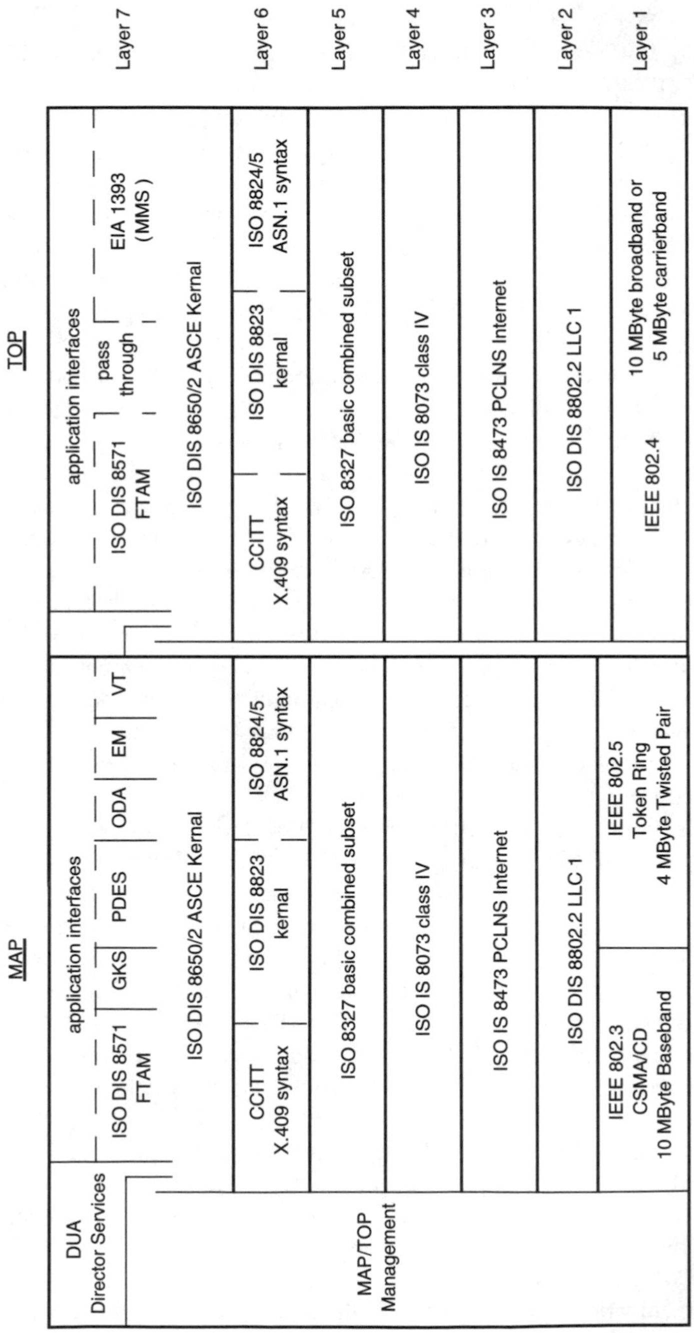


**FIGURE 2-17.** MAP and TOP standards.

The acceptance of MAP and TOP as the industry-wide communications standard will not be easy. The framework for the development of MAP/TOP is the ISO/OSI reference model. While the model defines the communication protocol in terms of a seven-layer structure and defines the functionality of each layer, it does not specify how this functionality should be implemented. Thus there have been many implementations of the ISO/OSI model particularly by the larger computer manufacturers. Existing proprietary standards, particularly from the major computer manufacturers, will not be easily replaced, and with this knowledge the systems designer would seem to be sentenced to a longer wait for OSI than ISO originally intended. The understanding and use of standards in this area by manufacturing systems designers can be critical to the long-term development of manufacturing systems. Standards such as MAP and TOP will ensure portability and compatibility among dissimilar computer-based systems and give an element of predictability to the way systems will ultimately perform. These standards will also lead to a rationalization of the way systems are integrated and ultimately lead to major reduction in data-integration costs.

## 2.5.2 Graphic Interface Standards

Graphic interface standards can be divided into two categories. The first are standards associated with the digital exchange of graphics between dissimilar graphics-based computer systems. The primary standards to emerge in this area are the CAD interfacing standards of IGES and EDIF. Both of these standards are already making the exchange of graphic information easier. However, they have limitations and are increasingly being seen as old technology. New initiatives are being developed in this area leading to the development of new expanded standards. PDES/STEP is now seen as the successor to IGES and EDIF.[63] PDES/STEP is being designed to expand the scope of IGES and EDIF to cover product as well as graphics data exchange between various graphics-based systems. Like MMS PDES/STEP, initiatives are being developed to fit into the ISO/OSI model and in particular the TOP standard. Together with GKS and ODA they form companion standards for the seventh layer of the TOP system architecture.[64, 65]

From the perspective of the manufacturing systems designer, the two standards that carry most significance for product data exchange are the established IGES and the emerging PDES/STEP standards. IGES is currently the most widely used of all data-exchange formats. The latest versions provide seamless data transfer between all CAD/CAM systems that have implemented the standard correctly. One of the difficulties experienced with IGES is ensuring that vendors have implemented the standard correctly in their pre- and postprocessor software versions. To ease this difficulty a number of centers have been established specifically for testing standard compatibility.

Like IGES, PDES evolved through the research projects initiated under the U.S. Air Force ICAM program. PDES has evolved from an earlier project entitled Product Data Definition Interface (PDDI). In both of the projects and in particular PDES the concept of simply encoding graphic data for exchange through pre- and postprocessors

(IGES) has been enlarged to include manufacturing data. Information such as process-specific data, tolerances, and datum positions are encoded into the preprocessed file. In its ultimate form the intention of PDES is to encode all the attributes of a product, including graphic representation, material types, color, process plans, and even history.

In Europe, the ESPRIT CAM-I project is carrying out similar work in the development of the Standard for the Exchange of Product Model Data (STEP). Currently both the PDES and STEP initiatives are being merged through joint collaborative efforts between the United States, Europe, and Japan. For the manufacturing system designer this means that when these standards become available and implemented by the many CAD/CAM vendors, the concepts of OSI can be extended into the office environment. Like MMS, PDES/STEP resides at the boundary between data integration and information integration. In this respect the manufacturing system designer will need to understand the impact of these standards and be able to use this knowledge for the development of application software for the manufacturing system.

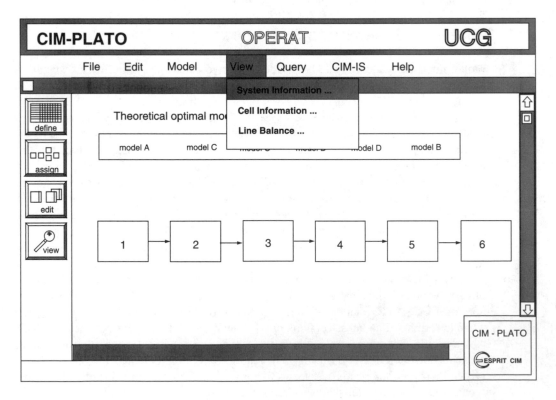

**FIGURE 2-18.**   User interface standard.

The second category for graphic interface standards and one which has become increasingly important for system designers to understand, is graphic user interface standards. Two of the most popular de facto standards in this area are the Finder interface standard from Apple and the Windows interface standard from Microsoft. Both standards have dominated software systems development for the PC level of computer applications. At the minicomputer level, standards such as OSF-Motif and X Windows are dominant. Figure 2-18 is an example of a manufacturing software system developed using the Apple Finder interface standard. The importance of systems designers appreciating user interface capabilities cannot be overstated. For example, in the development of supervisory software for a flexible manufacturing system, the software is often best visualized and specified through a series of simple drawings of the various user interfaces to be employed. If designers can create these drawings, software development groups are better able to understand requirements. In addition, when it comes to system proofing and acceptance, systems designers are better able to evaluate software deliverables.

### 2.5.3 Hardware Standards

In the classification presented earlier, hardware standards are those associated with physical equipment such as computers, controllers, and shop-floor equipment. Standards existing within this category are numerous. Many of the standards are de facto, having been developed by individual companies who have been responsible for the innovation and research of new hardware products. For example, in bar-coding equipment, the standard popularly used is one developed by a major bar-code equipment manufacturer. Intermec produces the most widely accepted de facto industry standard for bar code specifications. The greatest example of large companies achieving de facto industry standards is of course IBM with its Personal Computer. A sample list of hardware standards in this category is presented in Figure 2-19.

### 2.5.4 Standards Classification

The very large number of standards available make it very difficult to create a comprehensive classification that can be illustrated in one figure. However in Figure 2-19 a general classification is presented of standards that have been found useful in the implementation of the case studies that appear in Chapter 6. The classification is based on the categories presented earlier.

The section on standards is a general view of both existing and emerging standards. Standards are an important resource for the manufacturing system designer. Some standards allow the designer to create criteria for the acceptance of vendor hardware and software. They also set criteria for the efficient, safe operation of the final manufacturing system. Other standards are clearly evolving that specify the types of data and information that can flow between computer systems. Some of these standards attempt to bridge the gap between data integration and information integration.

## 2.6   OTHER TECHNICAL PLANNING TOOLS

This section examines a number of tools that are complementary to the modeling and system design tools described earlier. Although complementary, they are different in

| STANDARD CLASSIFICATION | | |
|---|---|---|
| CATEGORY | STANDARD | SHORT DESCRIPTION |
| COMMUNICATIONS (SOFTWARE) | MAP | Manufacturing Automation Protocol |
| | TOP | Technical Office Protocol |
| | DECNET | Digital Equipment International Network |
| | SNA | IBM's System Network Architecture |
| | KERMIT | Universal file transfer mechanism - public domain |
| | TCP/IP | Transmission Control Protocol/Internet Protocol |
| GRAPHIC INTERFACE | EDIF | Electronic Data Interface Format |
| | IGES | Initial Graphics Exchange Specification |
| | OSF-MOTIF | User Interface Design |
| | PDES | Product Data Exchange Specification |
| | STEP | Standard for Exchange of Product Model Data |
| HARDWARE | INTERMEC CODE 39 | Bar Code Interface Mechanism Standard |
| | RS 232 | Serial Data Transfer Standard |
| | RS 442 | Update to RS 232 |
| | IEEE 488 | Parallel Data Transfer Standard |
| | IBM-PC | Personal Computer Standard |
| | IEC 364 | Electrical Installation of Buildings |
| | BS 5940 | Specification for Design of Workstations and Offices |
| | IEEE 446 | Recommended Practice for Stand-by Power Systems |
| RELIABILITY | BS 4200 | Guide to Reliability of Electronic Equipment |
| | IEC 605 | Equipment Reliability Testing |
| | BS 5760 | Reliability of Systems Equipment and Components |
| SAFETY | BS 5304 | Code of Practice for Safeguarding of Machinery |
| | BS 5499 | Fire Safety Signs, Notices and Graphics Symbols |
| QUALITY | BS 5750 | Quality Systems Performance Criteria |
| | IS 9000 | Quality Systems Performance Criteria |
| | BS 5781 | Measurement and Calibration Systems Specification |
| PERFORMANCE MEASUREMENT | BS 5203 | Performance Measurement of Data Communications |
| | BS 6238 | Performance Measurement of Computer Systems |
| | MIPS | Benchmark Test (millions of instructions per second) for Computers |

**FIGURE 2-19.**   General standards classification.

that they assist designers in the operational details of specific manufacturing systems. In this regard they can be described as tools for the design of physical systems within operations. Before the emphasis on integration, these tools and their use typified the task of manufacturing systems design. For example, these tools are used for factory location, facilities layout, line balancing, material-flow optimization, and financial evaluation. While these tools are still important, the emphasis in this discussion is on tools that facilitate integration and in particular information integration.

Research into some of these more traditional tools has led to some interesting developments with regard to information integration. One project under the ESPRIT program is developing tools, which due to their design parameters are contributing to the information integration of manufacturing systems before they are installed into their final manufacturing environment. A review of this project here is seen as an important addition to the whole review of tools for integrated manufacturing systems design.

## 2.6.1 CIM System Planning Toolbox

The primary concept of the CIM Systems Planning Toolbox (CIM-PLATO) is that a set of integrated planning tools can be brought together to form a *toolbox* for systems design. A manufacturing systems designer may come to the toolbox, be assisted by it in choosing which tools he or she may need, and then use these tools in developing operational specifications. Because the tools are integrated, they share information with each other, and this facilitates shorter design times. This basic concept of the toolbox is illustrated in Figure 2-20.

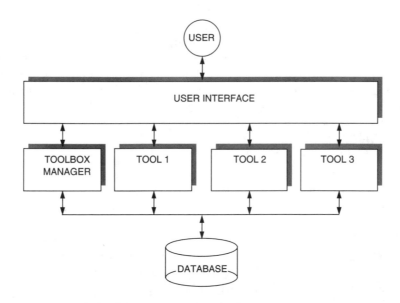

**FIGURE 2-20.**   CIM-PLATO Toolbox concept.

The concept allows the user access to a large number of tools through a common user interface, which then allows the user to experience a common *look and feel* to each of the tools he or she may decide to use. Selecting which tools to use is facilitated by a knowledge-based software called the toolbox manager. This software has metaknowledge on all of the tools available and is able to help the user to select tools through an interactive querying session. During this session the user may browse the database of the toolbox for more specific information. When the user has selected the tools, the toolbox manager configures the toolbox to allow only those tools that have been selected access to the processor and data-management facilities. The user can then use the configured toolbox for developing specifications for his or her particular system.

## 2.6.2 Flexible Assembly System (FAS) Configuration

To illustrate the type of activities that can be carried out in a typical encounter with the toolbox, consider the configuration given in Figure 2-21. In this configuration nine tools have been made available to the user for the development of specifications for a flexible assembly system. Typically, this type of system will contain a number of cells, each of which have tasks and associated resources. In the configuration, the user would first analyze the product for information about its components, their material, and any other special characteristics. He may then define the structure of the product and the tasks to be followed in its assembly. Once these tasks have been determined they can be assigned to cells along with the type of resource needed (e.g., robot). After this assignment the cells can be used for detailed material-flow planning and simulation to determine optimum buffer sizes and throughput rates. Then the FAS layout can be determined following financial evaluation of a number of alternatives. Finally, the new system can be analyzed from a control planning perspective so that once tested for motion, the source code can be generated automatically for the systems robots and PLCs. In all stages the user is enabled by these state-of-the-art planning tools that share information with each other to provide cost-effective, integrated manufacturing systems.

## 2.7    CONCLUSIONS

In this chapter we have discussed a number of the tools, techniques, methods, and standards that contribute to equipment, data, and information integration. One clear message given in this chapter is that there are many tools available to the manufacturing systems designer. Each tool provides a different perspective on manufacturing or what Checkland calls the "Weltanschauung" or implicit world view of the designer (Checkland's work will be described in the next chapter). This view can be software engineering oriented (as with ISP, SSADM, and EXPRESS), manufacturing engineering oriented (as with IDEFo and CIM-PLATO tools), or systems oriented (as with ISP and CIMPLAN). What is also clear is that the design of manufacturing systems in most cases involves the use of many of these views. Therefore, it is fair to suggest that in fact many of the technical planning systems or tools discussed need to be used in the design of a manufacturing system. The classification, presented at the end of the next chapter, indicates where each of these tools may be used. Chapter 3 will open with a review of some approaches and techniques that can be said to facilitate the other three elements of integration: management, system designer, and user integration.

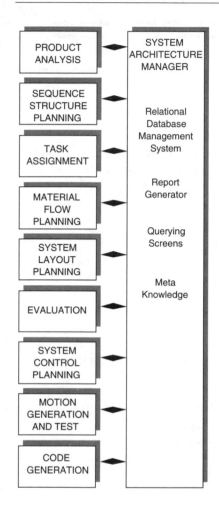

**FIGURE 2-21.** Toolbox configuration for FAS specifications.

# Social and Systems Planning

## 3.1 INTRODUCTION

In this chapter we will look at various social aspects of manufacturing systems design, which complements the emphasis on using good system development tools and methods discussed in Chapter 2. This was identified by Pava when he wrote: "A vast array of equipment is now coming to the market. Each task to which it is applied can be significantly enhanced. This technology will benefit most of the social sub-system of work - people and the organization of their tasks - complements the technical sub-system - tools and procedures."[10] This chapter is concerned with the social subsystem and the tools and techniques used for designing it. In the area of social subsystems, the *science of manufacturing* which was pursued by Harrington has to make way for less logical and less conclusive research. In the words of Kaplan: "(in the social sub-system) theory is more interpretative and experimentation less conclusive than it is in some of the physical sciences."[66] In this regard terms such as tools and techniques found in Chapter 2 are replaced with *softer* terms such as organization and learning. A number of important methodologies are identified for providing an approach to achieve social integration. These methodologies are sociotechnical design, the human infrastructure impact statement and the soft systems methodology. The inclusion of the soft systems methodology in this chapter is mainly for convenience, since this methodology addresses both technical and social design issues equally.

## 3.2 ORGANIZATION

According to Hayes "a company's organization is the glue that keeps manufacturing priorities in place and welds the manufacturing function into a competitive weapon."[22]

Every company needs to create an organization in order to breakdown to manageable levels the vast amount of activities that need to be performed. However, a dilemma exists between an organization that is divided into functions which create differentiation between functions, and an organization that must coordinate all of these functions including people and processes. This coordination is what was referred to earlier as social integration. Maintaining the balance between differentiation and integration is a balancing act with which most companies have difficulty in coping with. Building a purposeful organization is perhaps outside the scope of the manufacturing systems design function. However, in order to fully understand the issues involved in the design of manufacturing systems, an overview of organizational development is essential. An understanding of the concept of organizational learning is particularly important, as many of the poor results experienced by the installation of new systems can be attributed to poor learning capabilities on behalf of the users, managers, and subsequently the company as a whole.

## 3.2.1 Organization Structures

Many different types of organization structure exist, types of which suit some companies more than others. Ten types of structure or departmentation are common.[67] These are departmentation by simple numbers, time, function, territory, product, customer, market-orientation, process (or equipment), matrix organization, and strategic business unit. In manufacturing, departmentation by function, by matrix organization, and more recently by product and process are perhaps the most popular. Of these, departmentation by function as shown in Figure 3-1 is the most widely used.

The basic approach to functional departmentation is simple. Specialization is necessary in order to develop expertise. Hence, the various activities of the company are divided into functional areas as illustrated in this diagram, and staff and managers are

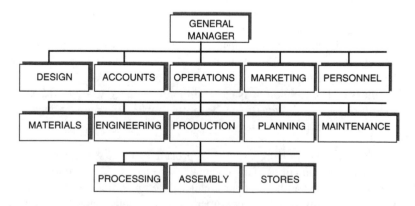

**FIGURE 3-1.** Departmentation by function.

assigned to each area. An organization hierarchy is then developed for each function. This creates a hierarchical structure within the organization. It also indicates the relative position of each function within the overall organization of the company. While this organization is still very typical of a manufacturing company, a number of organizational changes have occurred within this structure to create a hybrid organization consisting of attributes of all of the organizations referred to above.

The level of responsibility given to the departmentalized areas and the subsequent level of authority within each department is also important to understand. In its simplest form this means allocating responsibilities to departments and then providing the authority to those departments for carrying out these responsibilities. However, this is not as easy as it may at first seem. Two difficulties arise. The first is when authority falls short of responsibility. In this case any failures in a department cannot fully be blamed on the department because clearly some of the authority must have resided elsewhere. The second problem lies in performance measures for determining the effectiveness of each department's responsibilities. If these measures are not well defined, then effective management control information in not available. On the other hand, if measures are overdefined, they will miss cross-functional performance parameters which are now essential for effective systems development. So serious questions need to be asked. How does one evaluate a manufacturing organization on issues such as flexibility and order lead time? How are the responsibilities and authorities for these parameters distributed among the various functional departments involved? These questions have been answered to some extent and in various ways by many organizations. Some of the changes that organizations have gone through in addressing these kinds of issues will now be discussed.

As mentioned previously, a number of changes have occurred within the traditional functional organization to make it more responsive to the needs of the modern manufacturing environment. Nine of these changes have been documented by Majchrzak and are discussed briefly.[9]

*Adding Departments:* The addition of new departments is one way of coping with changes in this manufacturing environment without making radical changes to existing organizational structures. Perhaps the best example of this is the creation of a Quality Assurance Department in companies. This new department or function, which has been created out of the quality control function, has equal responsibility and authority with departments such as accounts, design, and operations. Another example in the context of advances in new technology is the creation of so-called CAD/CAM and CIM departments. The purpose of these departments is to steer other development departments and facilitate in the leading of project groups in the CIM technology areas.

*Common Reporting:* Another approach is to create a single department head for two closely related functions. The advantages are clear. Having one manager with increased responsibility and authority over a wider functional area means more cooperation among the people involved in the department, a better understanding of direct responsibilities, and a better sense of taking a task and completing it from start to finish. However, one problem is the level of responsibility on one person's shoulders and the ability of that person to manage effectively what are essentially two functions.

*Cross Functional Committees:* The use of committees and so-called *task forces* have been widely used when manufacturing is faced with new challenges. They are perhaps the most popular form of liaison between functional areas. However, although they allow people to become involved, it can be very time consuming, often wasting numerous members' valuable time. If not effectively controlled by a chairperson, they can be viewed by potential new members as a new burden.

*Product Grouping:* Another approach to the altering of traditional company organization is to departmentalize on the basis of product or market. This is illustrated in Figure 3-2. In this organization a more focused factory is created where departments are organized around products. Much of the principles employed in group technology are shared with this model, and it has received much attention from progressive companies in recent years.

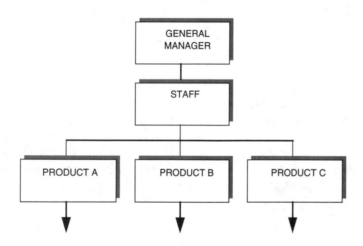

**FIGURE 3-2.** Departmentation by product/market.

*Process Grouping:* An alternative to product grouping is organization according to process. This is illustrated in Figure 3-3. In this organization common processes are grouped together. Three types of departmentalization are possible. (1) A plant is divided into its primary processes (e.g., Machining, Assembly, Stores, and Test). (2) Processes are divided between conventional and automated equipment, thus recognizing that both sets of technology are different and have different needs and require different staff. (3) Divide the processes into flow or parallel production. Thus a number of processes may exist. However, each process may carry out two or more classical process functions.

*Matrix Organization:* This organization allows people to participate and share responsibilities in a number of activities. People are assigned two types of responsibilities such as functional and product, committee and function, project and function, or project and process. To some extent this type of approach has been used extensively by the new quality-assurance departments, which require people to participate in quality committees as well as task forces.

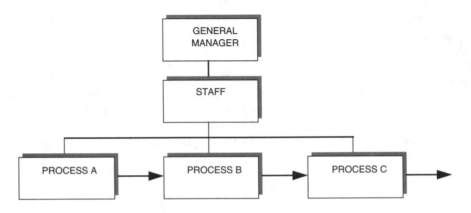

**FIGURE 3-3.** Departmentation by process.

*Decentralization:* The decentralization approach is used to delegate responsibilities down the management hierarchy in order to get more people involved in the decision-making responsibilities. The approach is very much based on the Japanese method of production management and total quality. In this method operators typically take on more responsibility for their work and in turn are given the authority for local decision making. Decentralization is very much dependent on management's perceived ability of operators. It is also dependent on management's ability to delegate both responsibility and authority to operators.

*Downsizing:* Another approach to creating a more responsible organization is downsizing. This involves cutting the size of manufacturing facilities to acceptable levels for the technology employed. In the CIM organization various limits have been set. Some corporations suggest a maximum of 500 employees for achieving the ideal organization. Downsizing is achieved by expanding capacity into other plants or by dividing one plant into two.

*Changing the Climate:* The final approach to the inherent barriers in the traditional functional organization is to create a climate that encourages coordination at many different levels. Coordination can be developed between operators at the shop-floor level, between levels in the hierarchy, and among the work force as a whole. This can be fostered through formal means (staff training and merit awards) and informal events (outings and social clubs). Clearly the way in which the climate is created will be a function of top management and its ability to lead by example in the issues that are important to the company as a whole.

Despite these approaches to the organizational dilemma that exists in most companies, barriers still exist between some departments. These barriers will probably remain until some radical changes are made to the management hierarchy. A key barrier is the lack of interplay between the functions of Manufacturing Systems Design and Management Information Systems (MIS) during systems design. In their work on production management systems, Browne et al. put forward some interesting ideas on this issue.[15] In general, the application of computers and computer-based systems in the manufacturing enterprise can be divided into two primary areas as indicated in Figure 3-4.

In this figure, applications normally associated with the shop-floor are placed into one area, whereas applications in the office environment are placed in the other. Each set of applications can clearly constitute a major project involving a number of disciplines and high expense. Some applications do not easily fall into either category. For example, an NC Part programming system may be either in the office environment beside design or on the shop-floor close to the CNC machines. A Flexible Manufacturing System, depending on its scope, may have some of its planning functions in the office environment, whereas the processing functions will clearly be on the shop floor. In short, these types of applications fall on the border between expertise and experience of the two design groups normally responsible for these two areas (i.e., MIS for office systems and manufacturing systems design for shop-floor systems). This departmentalization between the two systems design groups is now one of the primary focuses of manufacturing organizations for dealing with system designer integration.

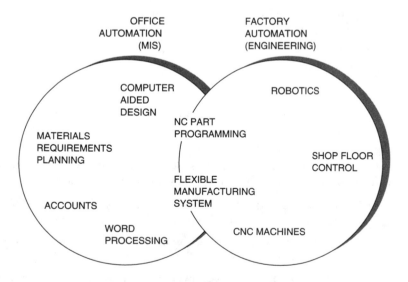

**FIGURE 3-4.** Computer-based Applications in Manufacturing.

## 3.3 LEARNING ORGANIZATION

An important concept that can be seen evolving in the literature is the concept of the Learning Organization. The term *Learning Organization* is used in a popular and influential book by Hayles et al. entitled, *Dynamic Manufacturing–Creating the Learning Organization*.[5] In their book the authors analyze the American manufacturing industry and describe the lessons they have learned from their analysis. They attribute many of the ills in American manufacturing to the inability to *learn and improve* across a wide range of manufacturing activities. They conclude: "Continual improvement is sought not through procedures and organizational approaches alone but also through mechanisms that facilitate both individual and organizational learning across projects." Pava devotes a large proportion of his book *Managing New Office Technology* to the same theme.[10] He promotes the use of the analysis technique *Sociotechnical Design* as a method that can facilitate "development of organizational learning and change." He remarks in a similar tone to Hayes: "When management fails to evoke organizational learning and change, even the most sophisticated (systems)... cannot realize substantial benefits for the enterprise." What is clear from these and other references is that organizational learning is important to the development of manufacturing industry.

### 3.3.1 Learning Organization Defined

Pedler et al. offer the following definition of a learning organization. [The learning organization is] "an organization which facilitates the learning of all of its members and continuously transforms itself."[13] They go on to emphasize that a company that spends heavily on training is not necessarily a learning organization. The learning company must also be concerned with the diffusion of learned information, with individual self-development, and with development of group learning as a whole. The learning company may even include extending the learning concept to the suppliers and customers of the company.

An attempt at classifying the learning organization is to develop a two-dimensional framework.[68] In the framework traditional versus learning centered management is placed on one axis of a graph. A second axis is labeled with individual development versus organizational development. In the framework, illustrated in Figure 3-5, the learning organization is firmly placed as one that promotes the highest level of organization transformation with a learning-centered management approach.

Pedler et al. see this classification as implying certain assumptions that can be made about a learning organization. They indicate that a learning company is one that:

1.  Has a climate in which individual members are encouraged to learn and develop their full potential.

2.  Extends this learning culture to include customers, suppliers, and other significant shareholders.

3.  Makes Human Resource Development strategy central to Business Policy.

4.  Is a continuous process for organizational transformation.

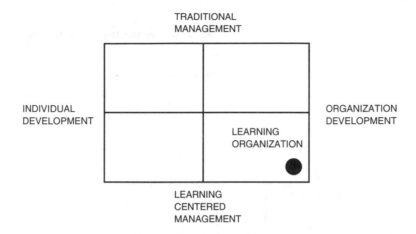

**FIGURE 3-5.** Classification of learning organization.

They make a clear distinction between what they call organization transformation and organization intervention. The latter they see as an activity involving what they describe as *outside intervention* or the use of outsiders for helping to carry out change in the organization. Organizational transformation on the other hand assumes that change can occur from within and generally without the need for outsiders.

## 3.3.2 Selected Approaches to Organization Learning

What is clear from the literature reviewed is that the concept of organization learning has been around for some time in one form or another. The terms *learning company* or *learning organization* are perhaps recent manifestations of what is an age-old quest. The first use of the term is attributed to Argyris and Schon, but the concept itself may be found is the earlier works on general systems theory.[69] Recent contributions to work on the learning organization are discussed in the next section.

Argyris and Schon in their work on organization learning differentiate between what they call single- and double-loop learning. Single-loop learning, they suggest, is what exists in many companies where an action is taken, the environment is monitored, and then a comparison is made with operating norms (Figure 3-6). In double-loop learning, which they claim is practiced in very few companies, an action is taken, the environment is monitored, a comparison is made with operating norms and then if necessary these norms are questioned and changed as required (refer to the thick line in Fig. 3-6).

They develop this model further to suggest that there is a third type of learning called deutero-learning, which is learning about learning. Single- and double-loop learning can only be achieved by the organization's individual members. However, in deutero-learning it is the organization as a whole which learns. They suggest: "When an organization engages in deutero-learning its members learn too about previous contexts for learning. They reflect on and enquire into previous episodes of organizational learn-

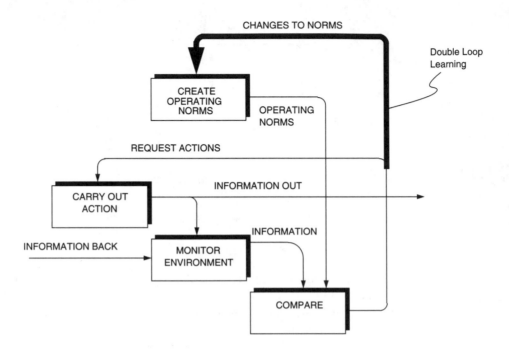

**FIGURE 3-6.** Single- and double-loop learning.

ing or failure to learn. They discover what they did that facilitated or inhibited learning, they invent new strategies for learning, and they evaluate and generalize what they have produced. The results become encoded in individual maps and images and are reflected in organization learning practice. Argyris and Schon go on to develop a methodology for expert facilitated intervention. The methodology involves the use of so-called experts in the concepts of the learning organization. This methodology involves four stages:

1. Mapping the organization's single-loop learning system.

2. Helping members of the organization make the transition from single-loop to double-loop learning.

3. Guiding and facilitating member's collaborative reflection on and restructuring of their own learning system.

4. Modeling and helping members to model good organizational dialectic in their efforts to detect and correct error in the organization's instrumental theory-in-use.

In his work on the transition to the learning organization, Revans proposes a different approach to the intervention.[70] His approach entitled "action-learning" views the manager as the only one who can possibly change the company's organization. Revans's approach can be described as individual centered, focusing on individual action. This is opposed to the group-centered approach of Argyris and Schon. Revans's focus on the individual is borne out in his so-called "qualities of autonomous learning systems." These qualities he suggests are as follows:

1. "Its chief executive places high among his or her own responsibilities that of developing the enterprise as a learning system: this he will achieve through his or her personal relations with his or her immediate subordinates.

2. The maximum authority for subordinates is to act within the field of their own known policies.

3. Codes of practice and other such regulations are to be seen as norms around which variations are deliberately encouraged as learning opportunities.

4. Any reference to what appears an intractable problem to a superior level should be accompanied by explanation, why it cannot be treated, where it seems to have arisen and a proposal to change.

5. Persons at all levels should be encouraged, with their immediate colleagues to make regular proposals for the study and reorganization of their own work systems."[5]

A completely different approach to organizational learning was developed by Hayes, Wheelwright, and Clark who, rather than treating the concept of the learning organization as a subject in its own right, use it to create a perspective on how they see the various functions in manufacturing operate.[5] In fact the term learning organization, although prominent in the title of the book (*Manufacturing Excellence–Creating the Learning Organization*), is rarely used again anywhere in the book. Their work looks at all of the main activities in the manufacturing organization from the perspective of the learning organization. One example they use of the concept of learning in action is in the development of projects within the manufacturing system.

They see two approaches to systems development. The first is based on what they describe as the conventional paradigm to development improvement. The second approach they describe is the learning-centered paradigm to development improvement. Using the conventional paradigm, companies that implement changes in the manufacturing environment expect project cycle times to continuously improve over time as a result of project managers' learning from experience (refer to the Expected Pattern in Fig. 3-7). In reality this is not the case. Managers experience little learning due to a number of factors. Hayes et al. cite a number of reasons for managers' inability to learn: "an urge to reassign key resources to the next project before this one is finished… the separation (physically, organizationally and psychologically) of different functional groups… natural resistance to change in any organization… and staff and systems support groups preference to 'fine-tune' the status quo."[5] In reality, they propose that the *actual pattern* of cycle time for projects is more or less constantly high. Projects begin to exhibit difficulties and drift upward (actual pattern shown in Fig. 3-7); sirens are then sounded; and all the stops are pulled out to get it back on schedule. In effect the project managers spend all of their time firefighting, and very little learning takes place.

The new paradigm in the opinion of Hayes et al. requires a very different approach. They support the concept of small, frequent changes to the manufacturing environment (learning pattern in Fig. 3-7). In this approach learning is achieved in every project undertaken. Some projects fail, while others succeed. Because projects are incrementally implemented, failure does not mean a major loss. In this simple example of how the concept of learning can be employed in a company, Hayes et al. essentially argue that adoption of the learning approach is first the responsibility of the individual (i.e., project managers) who must then through the organization as a whole try to foster a new approach. This approach leads to a new paradigm for systems development.

Pava offers a view of organizational learning based on his belief that sociotechnical design offers a methodology that fosters this approach.[10] He outlines four activities by which managers can promote organizational learning and change:

1. Re-define Management Responsibilities... to foster organizational learning and change... and greater involvement by managers in issues concerned with system design.

2. Set Forth a General Policy... that leads others to explore opportunities for organizational learning and change.

3. Support Decentralized Initiatives... avoiding change by edict.

4. Promote Organizational Design... setting priorities, providing resources and more importantly providing an approach or methodology for organizational learning and change.

The sociotechnical approach, in Pava's opinion, provides the right type of methodology for promoting organizational learning and change.

Frick and Riis outline an approach depicted in Figure 3-8 for manufacturing systems design which takes cognizance of organizational learning issues.[71]

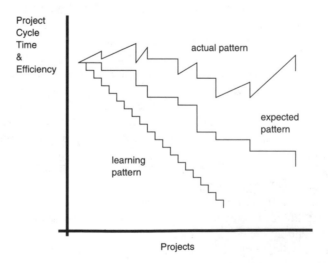

**FIGURE 3-7.** Approaches to project development.

**FIGURE 3-8.** Parallel change process for enterprise development.

These authors argue that organization learning is a means for achieving integrated and decentralized production systems. In one of their papers they describe a number of case studies that illustrate the use of organizational learning in parallel with other design approaches such as group technology. The concepts of organizational learning clearly support the concepts of the open system discussed in Chapter 1 and referred to many times in the review of the tools and methods of Chapter 2.

## 3.4  PROJECT ORGANIZATION

The composition and structure of the project team are critical to the management of the implementation of manufacturing projects. The composition of the project group is concerned with the selection of the project team members, whereas the structure is concerned with the management and information flow imposed on the team for effective communication and control.

In general, with current trends in the complexity of most manufacturing projects, members of the project group will and should represent a wide cross section of functions and disciplines. These will come from both the system design functions as well as the end-user community. They may also come from other support functions. Multiple disciplines are necessary because no single type of educational background provides the systems knowledge needed to understand the diverse impacts and implications of technologies such as flexible automation and other complex manufacturing projects. A large number of disciplines can be represented:

1. Manufacturing Systems Design
   a. Manufacturing Engineer
   b. Industrial Engineer
   c. Tooling Engineer
   d. Maintenance Engineer

2. Management Information Systems
   a. Systems Analyst

3. Product Design
   a. Design Engineer
   b. Design Draftsman

4. Operations
   a. Production Supervisor
   b. Production Planner
   c. NC Part Programmer
   d. Machine Operator

5. Other Support Functions
   a. Accountant
   b. Personnel Officer
   c. Quality Engineer

6. Management
   a. General Manager
   b. Functional Manager

7. External Functions
   a. Systems Consultant
   b. Organization Specialist
   c. Vendors/Suppliers

Clearly the potential number of the project members can be quite large. Thus the project group may be divided into a number of levels. The way in which this is done will depend on each individual project and the way in which it is being planned. However, a number of alternatives exist which are as follows.

*Two-Tier Membership:* This creates two levels of membership. One level is concerned with the steering or management level which could meet infrequently and mainly be concerned with status reports and strategic direction of the project. The second level is concerned with development or technical aspects of the project and would clearly meet more frequently and for more detailed discussion.

*Stage Membership:* This involves different people at different stages of the project and does not allow every member to expect to stay involved in all meetings. For example, design personnel could be involved during the initial specifications of machining resources, but later as the machining systems are being specified they could be excused from meeting attendance.

*Specialist Subgroups:* Another approach to getting as many people involved as possible is to form specialist subgroups with only the subgroup leader forming part of the overall project group.

The eventual approach to getting as many people involved as possible will be a direct function of the project scope and perspective of the project manager. Sometimes it may be necessary and even beneficial to involve an external facilitator to help the project group get started. The selection of group members should, where possible, be carried out by the project leader who should select members on the basis of motivation as well as perceived contribution. Vendor representatives are also useful, although not necessarily permanent members of a project group. A good systems supplier can bring invaluable expertise and knowledge to the project group even before contracts are signed. Finally, members experienced in the development of successful projects should be given a high priority when selecting project group members. However, such members may be hard to find. It is also possible that the perceived experience of one member in one area may turn out to be totally inadequate in another area and in fact hinder the development process altogether.

## 3.4.1 Project Group Structure

The efficient selection of project group members must be complemented by their efficient organization. A proper organization is necessary for the project to be lead effectively by the project manager and for members of the project team to get a sense of purpose and of belonging to the efforts of the project as a whole. In general, a large number of potential project organizations can arise. In the case where projects are being developed within one company, the following four types of organization are typical: (1) Functional Organization, (2) Lightweight Project Manager, (3) Heavyweight Project Manager, and (4) Tiger Team Organization.[5]

*Functional Organization:* The traditional functional organization is illustrated in Figure 3-9. This represents an organization where employees grouped by discipline report to a functional manager. The work between functions is coordinated by a set of agreed rules and procedures and communicated by the functional managers. The project is typically developed in phases. With the completion of each phase the project is handed from one functional group to the next.

*Lightweight Project Manager:* In this type of organization illustrated in Figure 3-10, the project is controlled by one project manager who uses a number of liaisons from the different functional groups. The group members report to their functional managers and simply cooperate with the project manager on project development issues. The project manager is usually a design engineer who has little effective control over the project members outside of their functional area.

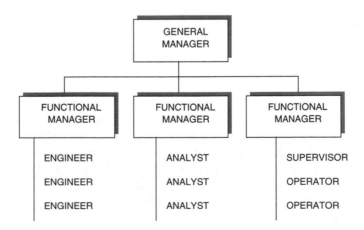

**FIGURE 3-9.** Functional organization.

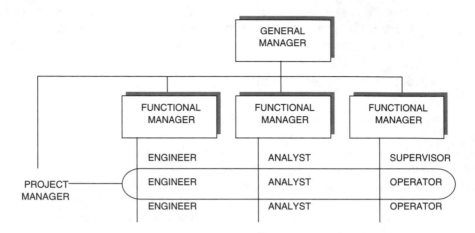

**FIGURE 3-10.** Lightweight project manager.

*Heavyweight Project Manager:* The third project organizational model involves what is known as a heavyweight project manager who has direct control over the members of the group. These in turn report directly to the project manager on project development issues and only to their functional managers on any other issues such as pay, conditions, overall work load, and career development. The project manager is usually a senior manager who is directly influenced by the expected benefits of the project. The project manager has experience, expertise, and organizational clout. The heavyweight project manager organization is illustrated in Figure 3-11.

*Tiger Team Organization:* The final type of project group organization involves a project manager and project members who are effectively removed from their functional responsibilities to form a tightly knit group for as long as is required to complete the project. The group is usually composed of experienced people familiar with an aggressive approach to project implementation and led by a project manager whose responsibility it is to execute the project using whatever resources are necessary. The tiger team organization is illustrated in Figure 3-12.

Each of these organizations has its strengths and weaknesses, depending on the company organization and project scope. However, in the development of *Next Generation Processes*, and *Single Department Upgrades* as defined in Section 1.4.1 the heavyweight project manager organization has clear advantages over the others. Heavyweight managers are usually clear about the overall company objectives of a project from a nonfunctional perspective. In their conclusion Hayes et al. recommend that the heavyweight project manager offers the best solution for the implementation of important projects.

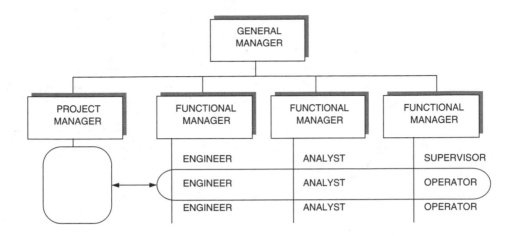

**FIGURE 3-11.** Heavyweight project manager.

## 3.4.2 Group Dynamics

The vast majority of manufacturing projects are implemented by groups of people, which are often made up of individuals from different disciplines and functional departments within the company. Selecting these individuals, who are often highly independent, is one of the greatest challenges for a project manager, who must try to ensure that the group becomes socially integrated before embarking on a major design exercise. He or she must be capable of creating a *dynamic* within the group which can overcome problems efficiently and thrive on success. As discussed earlier, the selection of the group (project) leader is one of the key decisions in implementing a manufacturing strategy. Once selected, a group leader must be aware of the factors that affect his or her ability in managing and controlling the work group. The following four issues are useful in helping group leaders understand the issues surrounding the creation of successful group dynamics: (1) group leadership, (2) group personalities, (3) group loyalties, and (4) group interaction.

### Group Leadership

In the previous section we saw that group organization has an important bearing on the success of a manufacturing project. In particular, we saw that the authority of the project leader is very much determined by his or her perceived role and position within the project group. Once in position, the style of leadership displayed by the project manager will have a direct influence on interactions between members of the group.

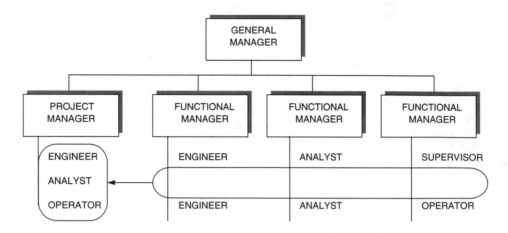

**FIGURE 3-12.** Tiger team organization.

Different styles of leadership may be required at different levels of project development. The project leader clearly has authority on issues such as scheduling, budget constraints, and resource management. He or she must also have mastered the appropriate skills to deal with these issues. An autocratic approach often requires a group leader to have strict discipline. On technical issues a different approach may be more appropriate. Actual leadership on technical issues may change among different members of the group a number of times during the life of a project. In this scenario, the role of the project leader becomes one of facilitator and administrator.

On issues related to technical design a more democratic style of leadership from a project leader is often the most successful.[72] Under this style, individual group members are allowed leadership over their own areas of expertise. The net result is a group of satisfied members who grow to adopt group goals and share group achievements. These observations are interesting in the context of manufacturing systems design where technology is changing rapidly. In this environment experienced group leaders are often technologically inferior to less experienced but more specialist group members.

## Group Personalities

The formation of a project group brings together people from different disciplines each with their own individual personality. Although these personalities often work well together, they sometimes clash, resulting in decreased efficiency for the group as a whole. In general three types of individual behavior are dominant in the group:

1.  A *task-oriented* individual is motivated by the work itself and its intellectual challenge. These individuals tend to be self-sufficient, resourceful, aloof, introverted, aggressive, competitive, and independent.

2.  A *self-oriented* individual is motivated by personal success. Technical tasks are often seen as a means of personal advancement. These individuals are often disagreeable, dogmatic, aggressive, competitive, introverted, and jealous.

**3.** An *interaction-oriented* individual is motivated by the presence and actions of other members of the group. These individuals are often passive, have low needs for autonomy and achievement, and are considerate and helpful.

Each type of behavior is exhibited by the same individual at different times. The general behavior of an individual can be attributed to one of these three. Groups consisting of entirely one type of individual are rare, and often group membership is constrained by the lack of choice. But where choice exists, groups composed of mainly interaction-oriented individuals are the most successful. In these groups a number of tasks will exist that require task-oriented individuals. These tasks are often easily "packaged," executed in isolation, and delivered efficiently.

### Group Loyalties

A well-led group induces individual loyalties to that group. This can be productive, particularly when projects are undergoing changes or problems. On the other hand, excessive loyalty can have its disadvantages, as for example when the group status quo becomes more important than the environment in which it operates. This leads to inflexibility in project changes and suggestions from sources outside the group. Another disadvantage of excessive group loyalty is when a new group member or project leader is *blocked out* by the other members of the group. To overcome this, new group leaders need to be very strong and assertive or perhaps be chosen from within the group itself. Another consequence of excessive group loyalty is "groupthink,"[73] which usually occurs under conditions of severe stress within the group. Under groupthink, groups sacrifice the benefits of various individual alternatives in favor of group decisions. The net effect is that decisions favored by the majority are adopted automatically without debate or consideration of alternatives.

### Group Interaction

Group interaction in general and communication between group members in particular are becoming an important aspect of successful project implementation. A number of factors influence the effectiveness of group interaction: the size of the workgroup, the workgroup structure, the status and personality of group members, and the physical work environment. The size of work groups can vary at different stages of project implementation. As size increases, interaction can become cumbersome and is usually facilitated through the group leader. Within the democratic style of leadership this is only one of a number of interaction mechanisms used. In large groups, opportunities may arise where specialist subgroups can be developed. These subgroups can have their own subgroup leaders who then speak on behalf of their subgroups at group meetings. In addition group leaders can promote individual group members to use specialist bilateral meetings between a small number of group members on specialist topics. In addition to meeting face-to-face, groups can adopt an environment of virtual meetings with each other through the use of techniques such as groupware, a term that has evolved to describe software systems that enhance the communication between members in a group. Another term used for the same concept is collaborative computing.

Groupware essentially provides an environment where groups can share ideas and information without the constraints of time and space. The most popular groupware tool is the mail facility available on most computer networks. Using up-to-date mail facilities, groups can be linked across rooms or even the globe through a virtual venue for meetings. Mail facilities allow each participant to work on the same data simultaneously. Other groupware systems have been categorized and include: Digitized Voice Applications, Project Management Applications, Natural Language Interfaces, and Group Decision Support Systems.[74]

## 3.5   SOCIAL AND SYSTEM DESIGN TOOLS

Although social systems are regarded as more experimental and less conclusive than technical systems, a number of useful approaches to the design of the social system have been developed. In this section three approaches that address social integration issues are reviewed. The inclusion of one of these tools, *Soft Systems Methodology*, in this section is merely for convenience. As we shall see, this tool deals with the concepts of systems and not solely the social subsystem. However, it is also very different in character to the technical systems design tools discussed in Chapter 2. The second tool discussed is Sociotechnical Design, which deals with both technical and social subsystems design. The third tool, Human Infrastructure Impact Statement, deals almost exclusively with social integration issues and how these issues can affect technical subsystem design.

### 3.5.1 Sociotechnical Design

The area of *job design* has traditionally been thought of solely in terms of the scientific management or the *Taylorist* approach to the design of jobs on the shop floor. While still widely practiced, it is increasingly being superseded by more human-centered approaches. These new approaches take cognizance of the changing social structure of the work force and the need to increase productivity by focusing attention on people rather than on machines. A large number of approaches are practiced that offer varying perspectives on the correct approach to job design. Such approaches include: Scientific Management, Job Enrichment, Job Rotation, Sociotechnical Design, Ergonomics, and Synergism.[9] Each approach offers a methodology for job design that has varying impact on the system being designed from the actions to be carried out to the design of work stations.

In recent years companies have been focusing on the technique of sociotechnical design as a way of gaining increased productivity from their operators, while at the same time creating a better working environment. This approach is of interest to this discussion as it has been developed on open systems concepts. It is this technique in particular that offers good approaches to the development of integrated manufacturing systems from the social perspective. It offers the possibilities of addressing the social integration issues introduced in Chapter 1.

## Sociotechnical Design Method

Sociotechnical Design (SD) is an approach or methodology that facilitates designers in the holistic analysis of social and technical subsystems within one organization or design project. It is described as a self-design method which allows designers to improve their own environment. It began as an approach to improving productivity in coal mines following World War II. A team of investigators headed by Eric Trist, from the Travistock Institute discovered a seemingly novel format of work organization in coal mines. This team discovered that for high performance, technology and work organization need to support each other, and to produce this, it may be necessary to underutilize equipment.

From the outset this caused a major debate in an environment that called for the increased use of mechanization and that often viewed people as a necessary evil in the work system. Subsequently, the legacy of the need for mechanization became the need for automation, and the notion of unmanned factories seemed to create the ultimate and ideal conclusion in designers' research and development efforts. More recently, however, most designers have come to realize that people are and perhaps always will be an essential part of the manufacturing system. They have realized that approaches to systems design must be better adapted to tackling social and technical systems analysis. SD provides such an approach, and when coupled with a number of other tools it provides a powerful resource kit to facilitate designers in achieving project goals. The SD methodology is well documented by Pava who worked with Trist on its development and documentation. We will follow the approach adopted in his book in the text that follows.[10]

Pava describes the SD methodology as consisting of two elements. The first is theoretical and contains various ideas about and ways of looking at systems. The second element is described as a procedure that sets out in five stages an implementation approach to the development of individual design projects.

## Main Ideas of Sociotechnical Design

The main ideas that underlie the SD approach to systems design can be divided into those that help us view and understand the operation of systems and those that mold the way we think about change in such systems. Below is a list of the more pertinent ideas in relation to the case study that follows.

*Systems Transform Inputs into Outputs:* The basic concept underlying this idea is that we can view systems (manufacturing systems, autotest systems, assembly systems, etc.) as comprising a set of inputs, a set of outputs, and between them a transformation process. The role of the transformation process is to produce desired outputs based on the inputs. On the face of it, this idea seems trivial; however, it suggests that we examine systems in terms of these three elements. Later we shall see, for example, that varying inputs (variants) are of primary importance in systems design and that the form of the variant can either control the transformation, be consumed by it, or simply be used as a resource. The form of the output also varies and can be either a desired result, a measure of performance, or a feedback to other transformation processes.

*Systems Regulate Activities through Feedback:* This idea suggests that feedback (information, results, requests, etc.) is used to regulate systems performance. Feedback is essentially information about mismatches between desired output and actual output; it is used to improve performance. Pava describes the idea of feedback in systems, as "systems capable of self-regulation, without excessive supervision, through the agency of shared goals." Self-control within the transformation process rather than between a transformation process and its environment is emphasized.

*Systems Are Open:* Another idea dominant in the theoretical concepts of SD is that systems are open. The concept of open systems introduces two environments in which the transformation process exists. The first is the transactional environment, in which the process is affected by and can have an effect on other processes in the environment. The second is the contextual environment, in which other activities can have an effect on the process. However, feedback from the process has no effect on other activities in the environment.

There are a number of other ideas associated with the theoretical foundation of SD, including concepts aimed at the approach of designers. Included in these are the ideas of self-design, minimum critical specifications, and open-ended design process. In self-design the concept is that organizations can only ensure efficient change if members within the organization are responsible for the change. The use of outside bodies such as consultants and research centers must be carefully managed to ensure that any change is channeled through the organization's own designers. In minimum critical specifications, SD promotes an approach to design that allows only the minimum critical specifications to be designed. Often designers focus on their own special interests and add features to a design that are not critical or even desirable, to the detriment of more critical features. Finally, the idea of an open-ended design process allows organizations to view their change process as ongoing, thus avoiding the creation of state-of-the-art showcases which are often expensive and out of date as soon as they are completed.

## Stages in Sociotechnical Design

The procedural part of SD consists of five stages, as illustrated in Figure 3-13. In this figure a further or initial stage is also shown *Entry*, *Sanction*, and *Start-up*. Pava describes this stage as occurring before SD begins. The stage designated as Stage 0 essentially entails preparing management and the design group on the workings and implications of the SD approach. Following this learning stage, the SD approach can be adopted confidently.

*Initial Scan:* The initial scan involves looking at the system in question and identifying the major variances, activities, feedbacks, and the system's mission or strategy. The initial scan can be viewed as a rough description of the system in terms of its sociotechnical factors. Close attention is paid to analyzing a system's philosophy and mission statements. Often mission statements are short and brief and to uncover their deeper meaning it is necessary to examine the views of various members of the enterprise. Quantifying or articulating the mission of a company can be difficult. Three approaches are recommend.[75] First, a mission when put into practice must make tangible differences. Second, it should be understood and accepted by members of the organization. Third, a mission statement should include criteria for future improvement.

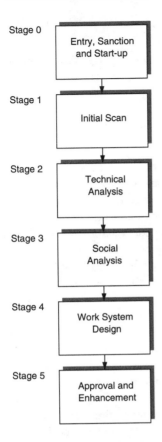

**FIGURE 3-13.** Stages in the sociotechnical design approach.

*Analyze Technical Subsystem:* This involves the detailed analysis of each of the activities in terms of which cause the main variances and how these variances occur. Variances cause unnecessary change in the process and lead to things going wrong. Three types of variance can be identified: (1) those that control or constrain the process, (2) those that are consumed by the process, and (3) those that are used as resources by the process. Typical activities include detailing the conversion processes, identification of links between processes, determining variances and their effects on upstream and downstream activities, and determining the levels of control necessary to avoid the effects of variances. Variance analysis requires impersonal and accurate detailing of variances. People must be encouraged to think about their organization in relatively objective terms, while simultaneously pinpointing the most important variances.

*Analyze Social Subsystem:* This stage involves the examination of the jobs affected by the system, how these jobs are coordinated, and how these jobs match the various psychological job criteria. As with the other stages, a systems view is taken where each environment, contextual and transactional, can have an impact on the target system. Psychological job criteria are discussed in Chapter 4. Issues to be addressed include division of labor, coordination, and fulfillment of psychological job criteria. The essential goal is to allow the design group to examine the social infrastructure with respect to its effect on the variances of the system.

*Proposals for Work Systems Design:* The next stage involves the generation of proposals based on information gathered in the preceding stages. The proposal will usually have three parts: (1) stipulate the mission of the work system, its inputs and outputs, its variances, and its responsibilities, (2) define the work organization that will form the best fit to the social and technical subsystems, and (3) stipulate technical upgrading to improve the control of variances for the productive execution of work.

*Approval and Enactment:* The final stage in the process involves obtaining approval from management for the proposed design and its enactment. Pava stresses that SD is primarily a process of analysis prior to implementation, and as such its main input to the design process is the provision of structured information to allow better decisions. The SD approach also allows for the unfreezing of long-standing assumptions and constraints. In turn, the participation in the process by the design group can create within the group so-called *champions of change*. These champions who have been awakened as a direct result of the SD process are then able to implement change along the self-design principles outlined earlier.

The sociotechnical design approach to systems development is both practical and systematic. It is clear that in order to gain more from systems, the social element must be considered as acting in concert with technical subsystems. The approach to the development of a new system from a technical viewpoint must include the examination of variances and feedback within the three environments in question: (1) the target environment, (2) the transactional environment, and (3) the contextual environment. The approach to development from the social viewpoint requires that each job in the system be examined with respect to the tasks performed, to the coordination of the jobs in order to provide efficient production, and to the satisfaction of psychological job criteria. The sociotechnical design approach is effective because it permits a marriage between two critical resources in the transformation system—technology and people.

## 3.5.2 Human Infrastructure Impact Statement

In her book on the human side of factory automation, Majchrzak proposes an interesting approach for the design of manufacturing systems.[9] This new approach addresses what she sees as the flawed method of the traditionally technically oriented design. She believes the technical approach to be characterized by six basic flaws:

1. There is an assumption that optimal technical utilization can be achieved primarily through technical factors. In reality technical factors are least likely to produce desired results. Majchrzak quotes an major study which concludes "the main stumbling blocks in the near future for implementation of programmable automation are not technical, but rather are the barriers of cost, organization of the factory, availability of appropriate skills and social effects of these technologies."[76]

2. Emphasis is placed on a reduction of direct labor costs as the rationale for new technology. This focus results in people being given second place in the design of operations for new systems. As mentioned briefly earlier, factories without people are impossible to achieve in most situations, and people will always remain an integral part of the operations of a factory. People will be required for complex assemblies, for maintenance, for control of machining systems, for diagnosis, and for tasks involving the interface between machining centers.

3. Even when social issues are addressed, the technical approach focuses on narrowly defined areas such as job descriptions and training. While these areas are important, issues such as organization, interaction among people of different functions, responsibility, and management often yield better results from design efforts.

4. Failure to preplan for the human infrastructure can create obstacles on the shop floor that may prevent or slow the utilization of new technology. These obstacles range from lack of motivation and unintentional resistance to sabotage and even to strikes. Off the floor, support staff are often not properly informed of changes, find they are suddenly responsible for new tasks, and sometimes find they are receiving instructions from other functional department heads.

5. The technical approach assumes that the human infrastructure can adjust to changes rapidly. For example, training costs and duration of the implementation of complex projects are often underestimated. In addition, the peak performance expected from these changes are often overestimated. In many projects, new technology takes considerably longer to *ramp up* to peak performance. Operators, supervisors, and managers find that they need more time to adjust to changes and to develop the inherent skills necessary to fine tune systems operation.

6. Failure to consider the human infrastructure can be directly linked to management's failure to understand value judgments embedded in the implementation and use of new technology. Majchrzak cites a number of studies that found that managers' value judgments about quality of work corresponded to the amount of control and discretion operators have with new technology. If managers and designers believe that operatives do not need job challenge, then the systems they sanction or design will not incorporate the necessary social details. Often it is these social details that cause a system to fail or to succeed.

Majchrzak's approach to addressing these six flaws is based on an *open systems approach* to human infrastructure planning. The traditional benefits of change—increased productivity, reduced scrap, improved utilization, better organization, etc.—can all be achieved through a process of change in the human infrastructure, she argues. In the open systems approach "the critical question is not what the components are, but what the nature of their interaction is."[77] In effect, an organization is composed of multiple components. The interrelationships among these components with respect to the change process determine organizational effectiveness.

## Human Infrastructure Components

The major components of the human infrastructure and their interrelationships with the change process are illustrated in Figure 3-14. In this model, three stages are shown that represent the waterflow effect that early decisions have on later options. These stages involve decisions about equipment, decisions about jobs, and decisions about various aspects of the organization. Each of the stages have constraints imposed, both from outside their own environment and from other stages within their environment. Each of the three stages takes place as a result of a change process. This change process itself has various characteristics that affect the results of the changes that occur.

*Equipment Decisions:* According to Majchrzak, the open systems approach to decisions regarding equipment and the human infrastructure can be based on six parameters: Integration, Reliability, Rigidity, Work Flow Unpredictability, Feedback, and Safety. By considering these parameters the best equipment for the interaction between technology and the human infrastructure can be found. Each parameter draws into the design process various issues that require decisions by the design group. In integration, the levels of integration among various functional islands will require analysis of responsibilities and task assignment.

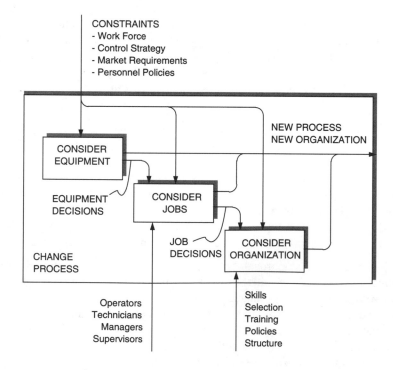

**FIGURE 3-14.** Main components of the impacts on human infrastructure.

Reliability of equipment, which by its nature is often part of a complex sociotechnical system, requires analysis on the degrees of interaction between people and machines. Often the mechanical part of a machining system has high reliability, but when coupled with other parts of the system, people, management, etc., the reliability is much reduced.

An important part of the performance of a large manufacturing system is rigidity or the flexibility of a system to make parts in a number of ways if one way breaks down. Systems need to be designed around the reliability of the total systems and not solely the mechanical element for which data are readily available.

Work flow unpredictability is a parameter based on the need for operators to have a degree of unpredictability in their work in order to find challenge and consequently to improve their performance. Designing unpredictability into the system is a more difficult task for designers than creating repetitive work flow systems.

Feedback is critical for manufacturing control and management. Often people are removed from the feedback loop because of new sensors, control equipment, etc. No feedback system is totally foolproof, and designers must ensure that operators also have the opportunity to become involved.

Finally, decisions about equipment safety are now critical, and the proliferation of safety standards in manufacturing facilitates the design process.

***Job Decisions:*** Decisions regarding the roles of individuals involved in a new system can be broken down into three categories: Operatives, Technical Staff, and Management. The design of operative jobs is often facilitated with social design tools such as ergonomics, time and motion studies, task analysis, and scientific management. Four specific dimensions to the design of operative jobs are identified by Majchrzak: (1) the coordination needs between operatives and other personnel, (2) the information needs of the operative, (3) redundancy checking by operatives of machine operations, and (4) the degrees of discretion that operatives have in their work environment.

Decisions regarding technical staff often revolve such tasks as programming, maintenance, quality and production control, accounting, and engineering. New systems often mean new ways for carrying out these tasks, new training needs, and better task assignment. Finally, decisions regarding management functions can be divided into two areas—management and supervision. System design can affect manager's planning and control activities over a new system. For supervisors decisions are often required regarding new job descriptions, new training and perhaps applying change to more highly skilled personnel.

***Organization Decisions:*** Decisions regarding the organization have been broken into five categories by Majchrzak: Skill Requirements, Selection, Training, Personnel Policies, and Structure. Skill requirements refer to the skills required by operators, managers, and technical staff in performing tasks. These skills include abilities in areas such as perception (concentration, attention, etc.), conception, dexterity, discretion, and human relations. Selections issues involve decisions on such topics as requirements, criteria for selection, people involved in selection, and source of potential staff (in-house or outside). Training decisions are made on issues such as courses, course content, amount of training,

training types (class room or on-the-job), and source of training expertise (in-house or out-side). The next category of organizational decision is personnel policy. Issues here include job descriptions, career paths, pay systems, job security, and labor relations. Decisions regarding Organizational Structure and in particular the creation of an organization that can learn and adapt to change have already been discussed earlier in this chapter.

*Constraints:* A major component of the open systems approach to design of the human infrastructure is the identification and analysis of various organizational constraints, which directly or indirectly affect decisions about change. Majchrzak identifies four basic types of constraint: Work Force Factors, Plant Control over Resources, Market-place Predictability, and Human Resource Philosophy. Work force factors identify a number of particular characteristics about the changing work environment, such as job challenge, trust, unionization, and level of skill. Plant control over resources yields three additional characteristics: plant degree of autonomy, supplier agreements, and labor-force mobility. The third source of constraint, Marketplace Predictability, requires decisions on issues such as entrance ability into new markets, performance gaps in existing system, and market growth. Finally, the effect of the human resource philosophy on the constraints affecting the change process can be categorized into decision-making styles, motivational models of the operator, and human resource priorities.

*Change Process:* The change process, in particular an approach for facilitating effective change, is essentially what this book is about. With respect to the human infra-structure or social subsystem, Majchrzak identifies four key issues that are important. The first is Resistance to Change, which recognizes the natural tendency for all people to resist changes in the status quo. The second issue is Counter Resistance, or measures that can soften the impact of change on the environment. Education, persuasion, participation, and empathy are all techniques useful in counteracting resistance to change. The third process addressed by Majchrzak is the importance of the project-team organization and manage-ment, in particular those involving users in the design process. The fourth key issue is the identification of implementation steps. Each of these issues are addressed specifically else-where in this book.

Majchrzak's in-depth and detailed analysis of the effects of the change process on the human infrastructure is extremely important within the context of the three social integration issues discussed in Section 1.5.1, particularly User Integration. The other two—Management Integration and System Designer Integration—can also benefit sig-nificantly from her work.

## HIIS Tool

In order to create an easy-to-follow procedure that systems designers can follow in designing the human infrastructure, Majchrzak developed the Human Infrastructure Impact Statement (HIIS), which is essentially a structured chart designed to allow designers to supply information about the system being designed. By filling in the infor-mation and invoking questions about the systems with respect to the human infrastructure, the HIIS tool yields insights into open systems design. The basic structure of the HIIS is given in Figure 3-15. In this figure the vertical axis represents each of the major components of the impacts of change of the human infrastructure (i.e., Equipment, Jobs, Organization, Change Process, Constraints, etc.). Along the horizontal axis four assessments are detailed for each component:

1.  **Existing Conditions:** What are existing organization structures, job descriptions, tasks per person, training needs, etc.?

2.  **Different Equipment Parameters:** What are the effects of equipment parameters on each element of the human infrastructure?

3.  **Organizational Constraints:** What effect do constraints have on equipment parameters, human infrastructure, etc.?

4.  **Impact of Other Elements:** How will equipment parameters directly influence information needs of operators, skill requirements, etc.?

The HIIS tool is an important contribution to the design of the social subsystem in manufacturing. It facilitates designers in the detailed analysis of the impact of technological decisions. When designers use the structured information presented in this analysis, many of the flaws in the more traditional technological approach can be overcome efficiently.

| Effects of Human Infrastructure | 1. Existing State | 2. Impact of Equip. Params. | 3. Constraints on Impacts | | | | 4. Impact on Other Elements |
|---|---|---|---|---|---|---|---|
| | | | Work Force | Plant Resource Control | Market Predictability | Mgt HR Philosophy | |
| I Machine Operator Job Activities | | | | | | | |
| II Technical Support Functions | | | | | | | |
| III Management Functions | | | | | | | |
| IV Supervisory Functions | | | | | | | |
| V Skill Requirements | | | | | | | |
| VI Selection | | | | | | | |
| VII Training | | | | | | | |
| VIII Personnel Policies | | | | | | | |
| IX Organizational Structure | | | | | | | |
| X Outcomes | | | | | | | |
| XI Planned Change Process | | | | | | | |

**FIGURE 3-15.** Human infrastructure impact statement (Source: Majchrzak, 1988, p42).

## 3.5.3 Soft Systems Methodology

In his work on systems theory, Checkland proposes a new *soft* approach to the development of systems.[17] The approach described in his book *Systems Thinking—Systems Practice* is popularly known as the Soft Systems Methodology and is described by Checkland as a "process of analysis during which a conceptual model of the whole system under analyzes is developed." Checkland states that his methodology is based on the belief that systems (manufacturing or otherwise) cannot be treated successfully using "hard" structured techniques such as operations research. In manufacturing systems, problems tend to be ill-structured so that analytical approaches are only useful to a certain point. Instead *a 'softer' approach is required*.

Checkland describes his soft systems methodology by the seven-stage process illustrated in Figure 3-16. In this figure the seven activities shown are divided into two kinds: real-world activities and systems thinking activities. The real-world activities represent a description of the system that is characteristically unstructured and often ambiguous. In this methodology real-world activities extract this description which is then translated into the *language of systems* by the so-called systems thinking activities. In the systems thinking world, structured and unambiguous analysis takes place around well-defined principles where information is then re-translated into the language of the real world where in turn it contributes to change.

*Stages 1 and 2 (Problem Situation):* During this stage an attempt is made to build up the richest possible picture of the *situation* without imposing a structure on the system.

*Stage 3 (Root Definition):* A viewpoint or number of viewpoints are selected that can bring about some improvement in the problem situation.

*Stage 4 (Conceptual Models):* This stage involves building conceptual models of the system that are relevant to the problem situation and for which a root definition has been defined.

*Stage 5 (Comparing):* Reality is compared to the conceptual model allowing the user to decide whether or not the model needs to be improved.

*Stages 6 and 7 (Implementation):* In the final stage *feasible and desirable* changes are implemented thereby creating some modification to the system. This change is created by what Checkland calls "purposeful" individuals.

The Checkland methodology is an elaborate one requiring the user to appreciate and fully understand the systems concepts exhaustively reviewed in his book. It can be essentially described as a *learning* approach where the users change systems by essentially finding out more about its special problems. Having found some unstructured problems they can then choose the right view or to use Checkland's preference "Weltanschauung" (or implicit world view). To help the system designers remember this view as well as some other concepts in the methodology, Checkland develops a mnemonic termed CATWOE, which stands for Client (the subsystem affected by the main activity), Agents (people who carry out the activities), Transformation (process carried out by the activity), Weltanschauung (perspective adopted by analyst), Ownership (that which controls the activity) and finally Environment (interactions with a wider environment).

This soft approach is also being pursued in the GRAI laboratory (developers of GRAI) through Pun, who views the approach to modeling of systems in much the same way as Checkland.[78] His approach entitled INtegrated COnceptual REferencing MOdel (INCOREMO) takes what he calls an anthropomorphic view of a manufacturing system. The view develops two separate models called the organic structure and the behavioral structure of the system. Pun argues that the approach is more logical and that by treating manufacturing systems like the human body, it can be examined (modeled and analyzed) in much the same way as medical practitioners examine living systems such as the human body. The soft systems approach in some ways disagrees with the more structured approach clearly adopted by Harrington through IDEFo. It is, however, interesting to note that in one reference a comparison between IDEFo and SSM found them to have many similarities.[79]

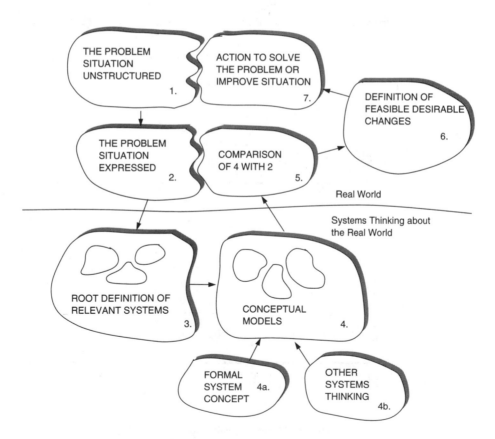

**FIGURE 3-16.** Stages in SSM (Checkland) Methodology (Source: Checkland, 1981, p163).

## 3.6   CLASSIFICATION OF PLANNING TOOLS

The classification of some of the more popular social and technical planning tools is presented in Figure 3.17. In this framework, the horizontal axis represents five types of modeling abstraction, which are as follows:

**Decision**—for modeling various organization and decision processes

**Process**—for modeling activities and process flow

**Data**—for modeling data requirements, storage, and manipulation

**Data flow**—for modeling flows of data

**Resource**—for modeling physical resources (e.g., optimum layouts)

Two areas presented in an earlier classification have been omitted. These are the User Interface and Process Logic categories. The process logic category has been redefined into the *process* category above. This is because process logic tools, although different from activity modeling tools, conveniently continue modeling of the manufacturing process after the activity modeling stage has been completed.

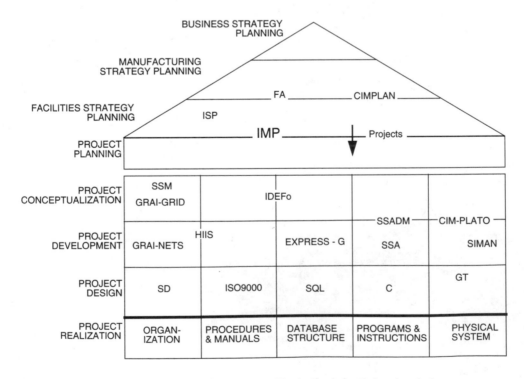

**FIGURE 3-17.**  General classification for design tools and methods.

The vertical axis in Fig. 3-17 represents the various levels of project management with the lowest row representing the target functionality. Concentrating on the top four levels, these are as follows:

**Business strategy planning.** This represents the activities associated with formulating a strategy for the entire business including financial planning, marketing planning, and manufacturing planning.

**Manufacturing strategy planning.** Discussed briefly earlier, this deals with translation of the business strategy into manufacturing related strategies.

**Facilities Strategy Planning.** This translates the manufacturing strategy into facilities related strategies.

**Project planning.** This represents those activities involved in the planning of individual projects which meet the objectives of the facilities strategy, and consequently, the manufacturing and business strategies.

The remaining project management levels from Project Conceptualization right down to Project Realization represent essentially varying degrees of *system model* for the manufacturing system. These models progressively define, in more detail, the specification and ultimately detailed operation of the manufacturing system.

In many of the boxes indicated in this framework, we have identified various popular methodologies and tools as useful and positive influences on the design process. Each of these methodologies facilitates a structured and open approach to social and technical systems design. In the area of *manufacturing strategy planning* and *facilities strategy planning* some tools exist that have only certain planning/design functions. References for most of the tools appear elsewhere in the text.

The IMP methodology, to be introduced in Chapter 4, is also located in Figure 3-17 as a technique to be applied in the areas of Manufacturing Strategy Planning, Facilities Strategy Planning and Project Planning. It is proposed that IMP represents a methodology which offers a more realistic breadth and depth of application to the systems analysis domain as a whole. IMP is also concerned, due to the various theoretical principles it adopts, with precisely what tools can and should be used in the lower levels of the classification in Figure 3-17.

## 3.7    CONCLUSIONS

In this chapter we looked at a number of areas that contribute to what was referred to in Section 1.5.1 as social integration. The concept of a learning organization viewed as a model through which much integration can take place both between people and between people and technology. We saw how the concept of the learning organization supports the open system approach to systems design. What also emerged is that the learning company movement, while not new, is gaining momentum.

We also discussed the organization of the project team, particularly from the learning organization perspective. It is clear from research that this special organization within the manufacturing enterprise is critical to successful manufacturing systems implementation. Issues such as structure, project manager selection, and group interaction were all seen as important prerequisites to project development.

For project development three specific methodologies were outlined which support open systems theory and claim to offer *logical* approaches to the design of Human Activity Systems. Each of these approaches not only deals with the social aspects of systems design but also with some technical aspects. Equally important each approach dealt with the relationship between social and technical subsystems.

# Theory of Manufacturing Systems Design

## 4.1 INTRODUCTION

In the preceding two chapters our discussion has mainly centered on developments in systems planning tools in the areas of technical integration and social integration. Together, these developments create a broad range of tools and methods available for the design of integrated manufacturing systems. Many of the areas discussed can be said to support the ideas and concepts of open systems to varying degrees. Selecting which of the many tools and methods to use in the design of manufacturing systems is one problem that designers face. In effect, the vast array of tools and techniques available creates its own type of complexity. Faced with such a wide selection, the designer may be forced to select one or more different tools and methods and because of this will adopt a particular view of manufacturing that employs the selected techniques. Many other issues perhaps more important to the design of the system may fall between the cracks. Another problem that arises from the development of so many approaches to systems design is that the concept of open systems can often become lost as the designer narrows his or her view into the perspective offered by any one of the tools. For example, the systems modeling tool, GRAI, is clearly built around open systems theory. The application of GRAI, however, is limited to only one specific problem domain (i.e., the decision flow process). Open systems theory tells us that systems are holistic and therefore examination of one subsystem in isolation from the others gives us an incomplete picture. GRAI and other methods such as IDEFo and SSADM have a lot to offer the systems analyst, but some sort of framework is required to place these and other tools in context with respect to (1) the theory of systems design and (2) where they are applied in the design cycle.

The methodologies discussed have been classified as either technically or socially oriented. Many of the tools discussed in Chapter 2 for example are exclusively technically oriented. In Chapter 3 each of the concepts and methodologies reviewed is described as being socially oriented. The use of sociotechnical design technique is clearly useful in designing sociotechnical interfaces, but is totally outperformed by say SSADM for the design of information systems. The problem here is not which tool to use but rather where and when it is appropriate to use *both* tools together.

One of the claims of general systems theory is that it gives designers a way of looking at systems that facilitates their understanding of the concepts of systems design. For many of the tools and techniques reviewed, this holistic way of looking at systems has been somewhat lost. If the benchmark created by GST is to be truly effective in providing designers with an open systems view of manufacturing, then the concepts and theory of open systems must take equal priority in the description of any methodology. This can be done either through guidelines or conceptual frameworks. The argument here is that system designers cannot avoid learning and understanding something about systems theory. Indeed, it could be argued that designers should be reminded of the principles of open systems design each time they embark on a new and often unique project.

These issues are compounded by the immediate reality of the system designer's need to decide how to plan, organize, develop, design, and integrate the new project into the social and technical environment. Hayes et al. recognized this need for developing expertise and knowledge in this area when they wrote: "If an organization develops skills in problem solving and conflict resolution, institutes new ways to organize projects, and develops broad-based manufacturing expertise, it can continually improve its development effectiveness."[4]

The intention of this book is to explore the issues outlined above and to create a new general systems methodology based firmly on the concepts of open systems. This general systems methodology will create a framework for the use of the many ideas, concepts, and techniques seen as necessary for developing the integrated manufacturing environment. In the survey of tools and techniques presented in earlier chapters no one particular methodology or technique was found that could allow the systems design group to explore systems design holistically. This is supported by Chiantella when he wrote: "The initial problem management faces is the lack of a methodology to explore CIM... at this writing there is not a set of management procedures to address the CIM opportunity."[8] This chapter introduces an approach that attempts to tackle these issues directly through what is called the Integrated Manufacturing Systems Design Procedure (IMP).

It must be clearly stated at the outset that IMP is not another modeling methodology. Rather, it is a methodology for systems design that attempts to synthesize a number of ideas and theories, systems design techniques, and modeling methodologies. A simplified illustration of this synthesis and an illustration of the development rationale behind IMP are given in Figure 4-1.

In Figure 4-1 a hierarchy is shown that illustrates the relative position of each of the main systems design techniques discussed in previous chapters. At the lower end of the hierarchy are system modeling tools that by and large consist of a technique and tool built upon various concepts. In the case of say IDEFo, its tool is the actigram that consists of boxes and arrows; its technique consists of procedures and rules used in constructing diagrams and text; finally, its concepts consist of structured analysis and open

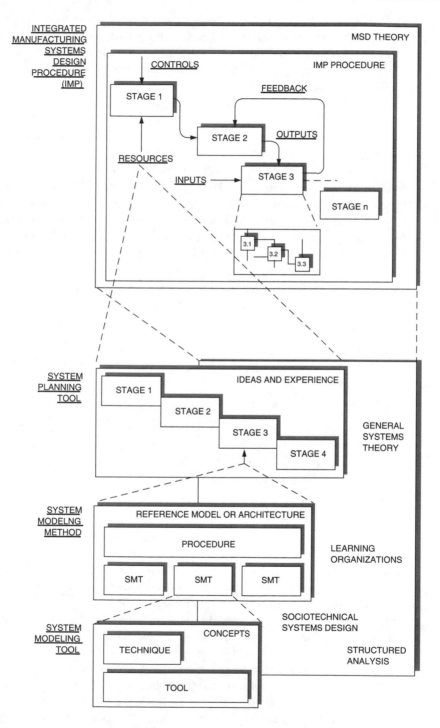

**FIGURE 4-1.** Hierarchy of planning systems.

systems theory. Many of the tools discussed have a similar construction, and many share a common conceptual base. The next level in the hierarchy of Figure 4-1 is systems modeling methods. In many of the methods reviewed in this book the methods are essentially a group of system modeling tools which are implemented according to a procedure and justified around some architecture or reference model. If we carry on the example of IDEFo, it can be clearly seen that the system modeling method of IDEF consists of three tools which are used sequentially—IDEFo, IDEF1x, and IDEF2. The procedure that guides us in the use and application of these three tools is contained in the IDEF specification, and all three find their justification in the ICAM architecture. In many of the system modeling methods reviewed, a reference model has been created to illustrate which certain tools should be used. In the GIM method, for example, the reference model identifies four domains of analysis in a manufacturing system—information systems, decision-making systems, functional systems, and physical systems. From this reference model the need for four tools that specialize in each domain has been identified: (1) Grai for modeling decision processes, (2) IDEFo for modeling functional characteristics, (3) IDEF1x for modeling data, and (4) a group of tools including group technology for modeling the physical system.

Moving up the hierarchy in Figure 4-1, we reach the system planning tool. In essence it is a tool built on experience and ideas about how systems should be implemented on a more strategic level. The tools can be said to consist of stages that are often sequential and sometimes parallel. Each of the stages can be implemented using some of the systems modeling methods and tools at the lower levels in the hierarchy. Also shown in Figure 4-1 is a box labeled with some of the more fundamental theoretical concepts reviewed so far, including sociotechnical systems design and general systems theory. These concepts do at times appear in the foundation of tools and methods. However, in general they are either lost or manifested through ideas and procedures taken by the authors of the tools. Indeed it can be said that for many tools and methods, users often remain unaware of their theoretical foundation. Consequently, when it comes to implementing or using a particular tool, its impact can be less efficient and often ambiguous.

At the highest level of Figure 4-1 is the basic concept for the construction of the integrated manufacturing systems design procedure (IMP), which in essence creates a marriage between many of the tools and techniques at the lower levels in the hierarchy. In addition, IMP attempts to clarify the theoretical foundations of manufacturing systems design. It is argued that by understanding this foundation the systems designer is better able to harness and utilize the individual tools and methods at his or her disposal. From Figure 4-1 IMP can be described as a procedure that lays out a theoretical baseline for systems design. Once understood by the designer, IMP then maps out in hierarchically decomposing stages, the implementation process for systems design. This process consists of activities, and within each activity the main controls, resources, inputs, outputs, and feedback can be shown. In this chapter we are concerned with outlining a theory of manufacturing systems design to facilitate the systems designer in understanding the manufacturing system in terms that ultimately help in its holistic design.

## 4.1.1 IMP Methodology

In Figure 4-2 the IMP methodology is illustrated as essentially consisting of two parts—a theoretical part and a procedural part. The theoretical part of IMP gets its main ideas from two major sources discussed earlier. The first of these is the theoretical, philosophical, and practical environment of manufacturing. This input includes various modern writings on both manufacturing systems design, writings in the general area of manufacturing systems, and the practical experience gained by various authors.

The existing social and technical planning tools are also an important input into IMP. The theoretical part of the IMP methodology may be said to be based on the latest thinking in the area of systems development and on key issues in the manufacturing environment in general. Concepts such as the learning organization, general systems theory, and structured analysis play an important part in the development of IMP. The theory is described in terms of seven principles of IMP which form the foundation upon which manufacturing systems are understood and the goals to which the IMP procedure must aspire.

The procedural part of the IMP methodology is described and detailed using the IMP model This model graphically displays, via hierarchical composition and activity modeling, all of the main activities in an idealized design procedure. The IMP model together with various tables, checklists, and charts provides a complex yet easily executed manufacturing systems design methodology. The procedural part of IMP is dealt with in Chapter 5. The theoretical part of the IMP methodology entitled IMP Theory can be divided into two parts, as illustrated in Figure 4-3.

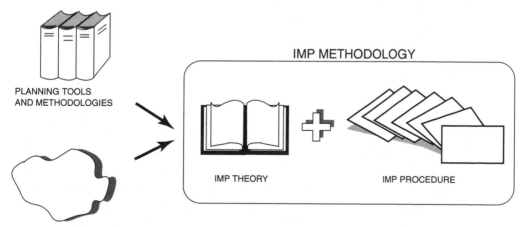

PLANNING TOOLS
AND METHODOLOGIES

IMP METHODOLOGY

IMP THEORY

IMP PROCEDURE

THEORETICAL, PHILOSOPHICAL, & PRACTICAL
ENVIRONMENT OF MANUFACTURING

**FIGURE 4-2.** IMP methodology.

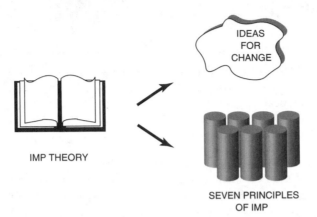

FIGURE 4-3. IMP theory.

## 4.2   IDEAS FOR CHANGE

In the chapters that have been presented so far, a number of areas have been discussed, and a number of key issues for the development process have been raised. These issues have been in a large number of development areas, from company learning to systems design tools and methods. In this section we will look at a number of *ideas for change*. These ideas can be seen a set of requirements for changes in the way manufacturing systems design operates. They essentially reflect the ideas created through the various commentaries given in the books, papers, and articles that have been reviewed. Some of the ideas also come from the author's own experience in a number of *real-life* applications which have been studied, some of which are presented in a later chapter. In the context of manufacturing systems design a number of important ideas for change arise. For clarity these ideas are classified under the general headings of the six integration types introduced earlier.

### 4.2.1 Ideas on Management

Management integration as outlined earlier concerns the social integration of managers within the manufacturing organization. Within this area a number of key ideas for change arise.

*Management Responsibilities:* The role of managers in the development process is critical to the success of the projects they implement. Managers are normally responsible for approving projects, following various presentations, and keeping track of project status. However, there is now a recognized need for managers to take a more active role in promoting organizational learning and change in order to extract the maximum benefit from new systems.

*Policy for Change:* Creating the right management environment ultimately leads to management having the capability to lead manufacturing activities more effectively. However, in order to communicate management thinking, there is a need for a policy statement for the organization as a whole to witness and to learn. A policy statement is required that addresses issues of strategy, mission, human resource policies, competitive advantages, operator training, and procedures. When encouraging change, management should also consider the extent to which change can be tolerated and to state precisely its policy with regard to resistance to change.

*Open-Ended Design Process:* Management should view the design process as open-ended. The manufacturing systems design process never really ends. There is a continuous cycle of change followed by learning, followed by change, and so on. The idea is primarily to facilitate organizational learning in management with respect to systems design.

## 4.2.2 Ideas on Designers

Designer integration concerns the social integration of designers involved in the design process. Ideas here are focused on the design team.

*Project Leader:* The role of the project leader in the systems design process is critical. Therefore the selection of the project leader for a particular project by management should be carried out very carefully and systematically. There is general agreement that project managers will be selected on the basis of their seniority and the amount of impact that the new project has on their area.

*Project Organization:* The organization of the project group for manufacturing systems design is generally considered to be arranged around the tiger-team approach outlined by Hayes.[22] Here the project leader has complete control over the actions of the group who report to him on all project-related issues. The group is comprised of users and designers.

*Conflict Resolution:* Conflicts always arise in project development. Some are healthy in that they bring attention to major issues, but most are unhealthy and arise from poor communication, a lack of clear objectives, organizational politics, or *carry over* from previous projects. Three rules of thumb are useful in resolving conflicts: (1) Bring to the surface the conflict as early as possible—through better communication, clearer objectives, and by promoting frank discussions. (2) Resolve conflicts through mutual accommodation—through a combination of bargaining, mutual agreement, and if necessary *hard* decisions. (3) Resolve conflicts at the lowest level of competency—conflicts between users and designers should be resolved within this group and should not necessarily involve intervention by top management.[22]

## 4.2.3 Ideas on Users

User integration is concerned with the social integration of users of manufacturing systems. Ideas here are more concerned with integrating users with the technical subsystem.

*Self design:* An idea basic to the design of manufacturing systems is that systems cannot be fully designed by outsiders. Systems users, be they shop-floor operators, supervisors, managers, or engineers are in the best position to offer ideas and alternatives on new systems. Members of the project group, such as systems analysts, project managers, and systems engineers, are best employed to act as facilitators to help the design team immerse itself in the project and to provide the project with leadership, variety generation, expert knowledge, and a liaison between individuals and groups in the project.[10] Another important role of the facilitator is that of translation—in particular translating user requirements into design specifications efficiently.

*Resistance to Change:* Resistance to change, from managers down to unskilled operators, can severely limit the success of new projects. Many of the causes of this resistance are attributed to lack of information concerning job security, job design changes, and new personnel policies. This resistance may be overcome in two primary ways.

**1.** Extensive communication delivered in a timely fashion about the new project to all affected personnel is essential. This may be achieved through training, orientation programs, and periodic information exchanges.

**2.** Involving users during the project design and implementation stages is important. This may be achieved through user participants on the project team, direct consultation with users on the shop floor, and consensus between operators and the project group about key design factors.

## 4.2.4 Ideas on Information

Information integration is concerned with the technical integration of information to facilitate the effective management and control of manufacturing systems. Ideas in this category reflect the need for structured approaches to information systems design.

*Structured Methodologies:* One idea on information integration is the use of structured methodologies based on open systems theory to design and analyze information flows in manufacturing. A large number of these structured methodologies exist to provide the design group with an unambiguous and structured approach to the analysis of information systems. In the preceding chapters it was shown that these methodologies could be applied to a number of areas of analysis, from functional through data structure and on to software code generation.

*Abstraction:* Another idea that is closely related to work in the development of systems architectures is that the systems can be modeled and described in terms of a number of views. In Chapter 2, for example, five views or levels of abstraction were presented for classifying modeling tools. Namely, decision, process, data structure, data flow, and resource. Other ideas include user-interface and process logic. Identifying the levels of abstraction in information integration and applying tools for analysis of these levels are important for overall systems integration.

## 4.2.5 Ideas on Data

Data integration is concerned with either electronic or manual communication between various subsystems. Ideas on data integration are more difficult to identify since most data integration issues are covered by developments in standards particularly with respect to communications networks.

*Standards:* One idea is the strict use of and adherence to standards, which are very important for promoting industry-wide conformance to common data specifications. Communications standards such as MAP, TOP, and RS232, as well as product-data-exchange standards such as IGES, PDES/STEP, and EDIF promote data integration of multivendor equipment. As shown by General Motors, manufacturing systems designers can exert sufficient pressure on manufacturers of computer products to make standards evolve more quickly. Once specified, these standards reduce data integration costs for everyone.

*Data Management:* A number of key ideas arise on issues to do with the handling and management of data. One idea is that all data should originate from one prime author. Any possibility of using duplicated or modified data should be avoided. In addition, data should be updated regularly from its source. Modification of data on reports, copies, or derivations of the original data is also to be avoided. In general, data should originate from mechanized procedures, such as from shop-floor control equipment. Finally, any data used in facilitating management decisions should be available in summary format and collated from the different original sources. Data belonging to separate functions should be held on physically different files or databases.

## 4.2.6 Ideas on Equipment

Equipment integration is concerned with the physical integration of equipment within the manufacturing environment. Like data integration, ideas in this area are mainly limited to issues in standards usage.

*Group Technology:* One idea that can be said to have become increasingly important in recent years is the use of techniques such as Group Technology for equipment layout. It is widely argued that all manufacturing environments, no matter how diverse or varied, can benefit significantly from the use of group technology techniques.

*Machine Redundancy:* From time to time machines break down and computer disks *crash*. An often critical idea sometimes forgotten by systems designers is to allow for machine redundancy. Machining systems, handling systems, computers, and computer networks all need to be designed with redundancy in mind. Often a balance is required between the additional cost of an extra AGV or computer hard disk and the chances of system failure. In critical applications this additional cost must be a part of the overall solution.

A large number of other ideas may be included here. The aim of these ideas is to help guide systems designers toward good systems design practice. The intention is that these ideas will always represent state-of-the-art thinking in systems design. Therefore, it is acknowledged that the ideas presented will change during the normal learning process. For example, it is likely that new, more important ideas may arise over time. These ideas will be included in IMP. Older ideas, which may be of a lesser priority, may for the sake of clarity be excluded from the IMP methodology. This concept of changing IMP over time is in line with open systems thinking. IMP like manufacturing is a type of system and therefore subject to the same learning traits.

## 4.3    PRINCIPLES OF IMP

The IMP procedure which is the *implementable* part of the IMP methodology is developed on seven key principles (Fig. 4-4), which can be said to be the pillars upon which the IMP procedure is built, and which make statements about manufacturing systems that give the IMP procedure a set of goals or objectives it must achieve. For example, if it is generally agreed that manufacturing systems must operate according to some manufacturing strategy or set of goals, then this must be reflected in any procedure that is being designed to augment the development of such systems. All of the IMP principles are derived either directly or indirectly from a number of already well-known methodologies and ideas used in open systems design. In particular, some of the concepts developed in General Systems Theory, Sociotechnical Design, and Structured Analysis are used regularly.

The creation of these principles is primarily to provide the manufacturing systems designer with a way for viewing manufacturing systems that he can refer back to again and again. It is acknowledged that IMP, not unlike many other methodologies, is in itself a system, and is influenced by changes in its environment. Therefore while the term *principle* evokes an air of permanency, it is accepted by the author that these principles may change over the evolution of IMP.

While each principle individually provides a useful way of looking at manufacturing systems, maximum benefit can only be achieved when each is used to complement the other. Definitions of each principle overlap to some degree the definitions of others. In fact, new principles can be conceived when concentrating closely on two or more of these principles in isolation from the others. Therefore, it is important to understand that while each principle can be applied individually, the principles are also connected together to form a comprehensive set of guiding principles for the way that IMP views manufacturing systems.

### 4.3.1 Principle #1: Manufacturing Systems are Goal Seeking

The first IMP principle states that manufacturing systems are goal seeking, which is an idea taken directly from general systems theory. All activities carried out in manufacturing are directed towards some goal or set of goals. The goals of manufacturing dif-

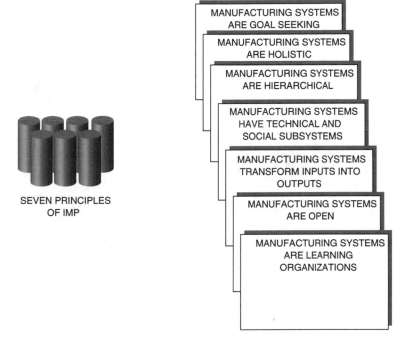

**FIGURE 4-4.** Seven principles of IMP.

fer, depending on the level of activity that is being examined. For example, the goal of an operations manager might be to get a quality product to the customer on time every time. On the shop floor the goal of a machine operator might be to produce component piece parts to specification. No matter which level of activity is being discussed, each goal will essentially be linked to another in order to reflect the overall goals of the company. In previous sections the concept of manufacturing strategy was introduced as the primary goal-setting activity within manufacturing. In terms of this first principle all activities both developmental and operational must have goals that adhere and contribute to a manufacturing strategy which in turn adheres and contributes to an overall business strategy.

The first principle also recognizes that these goals can be achieved in a number of ways. For example, the strategy that one manager might create may differ considerably from that of another manager. However, both strategies could conceivably result in the same system. This acknowledgment means that the IMP procedure must allow enough flexibility to allow managers to achieve the goals that are important. The principle of goal seeking and the links between goals in different functions at different levels in the organization is depicted in Figure 4-5.

In this figure goals are seen as covering the whole spectrum of management and operation of a manufacturing system, from strategic goals at the management level to goals expressed as methods and routines at the operational level. Using this principle

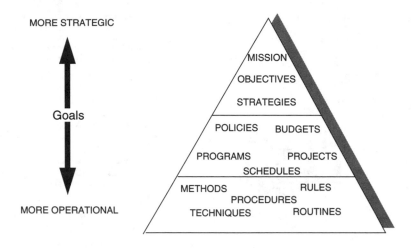

**FIGURE 4-5.** Hierarchy of goals.

the IMP procedure must be capable of acknowledging that goals are the primary controlling effect on both developmental and operational manufacturing activities. Also, as will be discussed later, the management of these goals must be complemented through *feedback* that can be used to assess whether or not the goals are being achieved and to what extent.

Goals may be outer-directed or inner-directed. Outer-directed goals are aimed at satisfying environmental ends whereas inner-directed goals are aimed at maintaining, perpetuating, and developing the manufacturing system. Outer-directed goals are primarily concerned with market-sector penetration and increasing sales. Inner-directed goals are primarily concerned with changing the manufacturing, design, and order flow systems to meet external goals.

## 4.3.2 Principle #2: Manufacturing Systems are Holistic

The second principle of IMP states that a manufacturing system is holistic, that is, it is an entity with interdependent and inseparable parts. Hence, to study one part of the system it is necessary to study, to some degree, the whole. There are various parts to a manufacturing system that interact with each other in various ways. If one dissects a system, interactions crucial to understanding are lost. The system must first of all be viewed as a whole, then progressively viewed more towards the perspective of the target subsystem.

A number of issues come into play when examining a system holistically. This is particularly true in the case of complex systems such as manufacturing. In previous sections a number of methods were discussed as representing manufacturing systems. Systems architectures were discussed as useful generic forms of the manufacturing environment. Systems design tools or structured methodologies were also introduced as useful ways for allowing system designers to create their own system architectures in a holistic fashion. In the design of systems it is important that similar tools for modeling and analysis be adopted which support the need to represent systems holistically. While holism is important for allowing the manufacturing systems designer a broad perspective on the system under change, equally important are three additional ideas introduced through the concept of structuring. These are seen as important in realistically curtailing of the amount of knowledge required to analyze the new system. These three ideas are *bounding the context*, *limitation of information*, and *viewpoint* (described in Softech[80]). Each idea essentially encourages the narrowing of perspective as soon as the target system and target parameters have been identified.

***Bounding the Context:*** While a holistic view of manufacturing systems is important for giving the designer an appreciation of how the various subsystems interact with each other, it is impossible for the design group to analyze everything due to budget, time, or technical constraints. The scope of the project from both a project management (developmental) and operational perspective needs to be defined and detailed while at the same time keeping the holistic nature of systems. Scope defines activities that can be said to be inside and outside the target subsystem. For example, if a system can be represented as comprising a number of subsystems as illustrated in Figure 4-6, then clearly a holistic view of the system within this level of abstraction is available to the system designer. This enables the designer to create a boundary around those subsystems that are important to the scope of the project in the knowledge that he/she also understands the subsystems that have been left outside the boundary. Subsystems inside the boundary are said to form the context for the project.

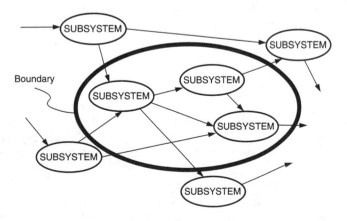

**FIGURE 4-6.** Bounding the context.

*Limitation of Information:* While stating that systems need to be viewed holistically, the amount of information that can be handled in any one instance by individuals in the design group is limited. For this reason the way in which the information is presented is important. All the activities in a manufacturing system could for example be illustrated in one figure, the usefulness of which would depend on the amount of information displayed. The more detail involved, the lesser the likelihood that the human mind will be able to comprehend the issues holistically. The more general the information, the better the chance of its being understood albeit at a more general level.

*Viewpoint:* Viewing a system holistically implies that an infinite variety of perspectives can be applied to any model representing the system. It is highly improbable that different individuals will agree on the accuracy of each view. Designers tend to agree only with their own views. Emphasis should be placed on all designers in the design group sharing a common view or way of looking at the system. Many of the tools that were discussed in Chapter 2 can promote this sharing of perspective. In summary, holism in IMP is about system designers taking a broad view of the entire manufacturing system prior to taking decisions on detailed systems design.

## 4.3.3 Principle #3: Manufacturing Systems are Hierarchical

The third IMP principle asserts that manufacturing systems are hierarchical. That is to say, manufacturing systems can be broken into subparts which behave differently to other subparts, but which contribute to the attainment of common goals. The hierarchical decomposition of major parts into smaller parts also explodes the goals and their horizons. Hierarchies exist in a number of manufacturing subsystems both social and technical. For example, the personnel hierarchy of a particular manufacturing company begins with the general manager at the top, perhaps functional managers at the next lower level, and right down to perhaps shop-floor operators at the lowest level. Another example is the now much-used production management hierarchy shown in Figure 4-7, which illustrates the hierarchy of functions within the product-order flow cycle. This figure shows the planning horizon of the production management function from strategic planning right down to operational planning and real-time control.

Hierarchy can best be represented by the technique of hierarchical decomposition widely used in many of the CIM Design Tools discussed earlier. The approach of hierarchical decomposition is to first model the upper layers in the hierarchy. Once this has been achieved, the upper level of the system is divided and a further more detailed description is applied to the resulting subsystems. These subsystems in turn are divided and so on until the levels of detail required are achieved. In Figure 4-8 an example of hierarchical decomposition is illustrated.

As with the previous IMP principle, the concept of hierarchy indicates continuous hierarchical decomposition, and this must be in some sense curtailed by the project group for the same reasons given earlier (i.e., Schedule, Budget, and Scope constraints). Two additional ideas arise; these are *levels of detail* and *levels of abstraction*. These are again introduced through the concept of structuring and seen as important in curtailing the amount of knowledge required to analyze the new system.

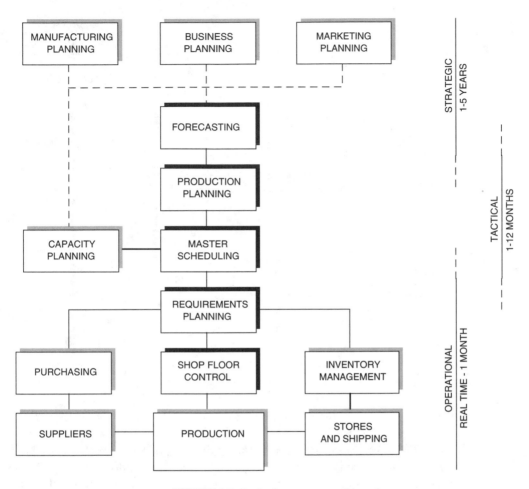

**FIGURE 4-7.** Production management hierarchy.

*Levels of Detail:* The idea on limitations of detail is closely linked with the idea introduced in the previous IMP principle—*limitation of information.* Essentially the idea is that the hierarchical decomposition of detail should only take place in areas that are within the scope of the project. For example, in Figure 4-8 not all of the subsystems in the second level of the hierarchy are illustrated as being further decomposed. In many design projects some subsystems at a relatively high level in the hierarchy are left unexplored. Further decomposition of the subsystem would either be excessively resource intensive or simply not useful enough for the creation of an overall understanding of the system. However, some areas are continuously decomposed, displaying ever-increasing levels of detail that when analyzed provide essential information.

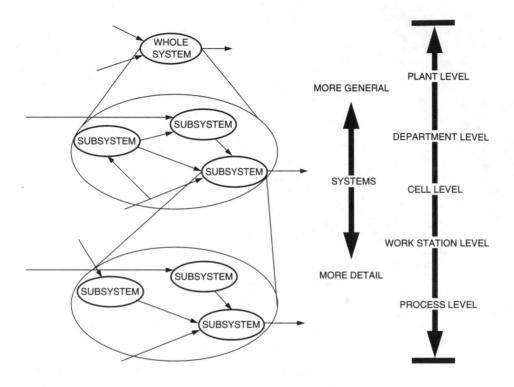

**FIGURE 4-8.** Hierarchy and hierarchical decomposition.

*Levels of Abstraction:* The idea of abstraction as applied to hierarchy states that various levels of abstraction exist which in turn create their own individual hierarchies and levels of detail. For example, a flexible manufacturing system can be described and modeled in terms of its order flow, product data flow, computer hierarchy, personnel hierarchy, and so on. In fact, a large number of levels can exist, and while each level has its own holistic and hierarchical principles they each form part of an even larger whole and hierarchy.

## 4.3.4  Principle #4: Manufacturing Systems have Technical and Social Subsystems

The fourth principle states that a manufacturing system consists of both social and technical subsystems. To some extent this is covered under the *levels of abstraction* discussed in the previous principle, but the emphasis here is placed particularly on the social subsystem. The principle is derived primarily from the sociotechnical approach described in Chapter 3, but with modifications to take cognizance of the various types of integration presented earlier in Chapter 1.

Social subsystems consume social energy in the same way that technical subsystems consume electricity, oil, and so on. While energy consumption can sometimes be taken for granted in the technical subsystem, it is of critical importance in the social subsystem. The basic forms of social energies are assertiveness, spirit, leadership, and adaptability. We can say these consume energy since they require exertion beyond simple survival. Creating the correct environment in order to foster this energy exertion is the primary role of the designer. The understanding of the technical and social subsystem in the context of this principle is best understood through the ideas of sociotechnical design as documented by Pava.[10] We will use Pava's description here for clarity.

*Technical Subsystem:* The technical subsystem consists of tools and techniques that convert input into output. The way in which the desired output reflects the actual output is controlled by variances in the flow of inputs and outputs. Variances are aspects of the conversion process which can go awry. Upstream errors that cause error downstream in the conversion, or target process, are called transmitted variances. Major activities outside the target system that can disrupt its performance are called boundary variances. These variances and their effect are illustrated in Figure 4-9.

The sociotechnical approach attempts to analyze systematic interrelations between activities and to monitor variances directly. The intention is to place the needed capability at the location of the variance. It involves placing information, authority, and the skills needed to control the variances where they occur before an error is created.

*Social Subsystem:* The approach to the social subsystem is more theoretical and interpretative because of the inconclusive nature of social theory and behavior. However, sufficient agreement exists to allow definition of the social subsystem as comprising two main issues: (1) division of labor and (2) methods of coordination used to control variances in the conversion process. The division of labor is concerned with jobs, that is, the tasks that people perform and the roles that people are assigned to fulfill. Once the tasks and roles are assigned, the jobs must be coordinated in order that they will be productive. The most common method of coordination within organizations is the setting out of rules and regulations whether written or implied. During the design process the sociotechnical approach can be used to look for new ways of grouping tasks and coordinating work flow to satisfy the social needs of the operators. These needs can be examined through what Pava calls the "Psychological Job Criteria," which if analyzed properly "can evoke sustained commitment and involvement from the human side of the system." Pava outlines six criteria:

**1.** "Autonomy and discretion. Engaging work provides a blend of opportunities for responsibility and self-management with a balance of clear guidelines for action.

**2.** Opportunity to learn and continue learning on the job. Engaging work provides ample opportunity for acquiring knowledge and skills.

**3.** Optimal variety. Work that engages people allows them to pursue an optimal variety of activities.

**4.** Opportunity to exchange help and respect. Work that engages people creates conditions in which fellow operators can and do exchange help and respect.

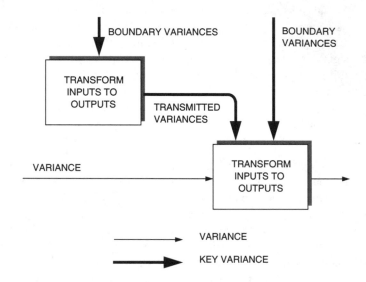

**FIGURE 4-9.** Transmitted and boundary variances.

**5.** Sense of meaningful contribution. Engaging work provides operators with a sense that their contributions are worthwhile in that they respect a challenge successfully met and they contribute to society.

**6.** Prospect of a meaningful future. Engaging work promises advancement, which encourages personal growth, and offers appropriately higher compensation."

In the fiercely competitive environment where new technologies are continuously being introduced into the manufacturing system, this principle tries to encourage the realization that there are inherent positive facets in one of the most important resources in the company—people.

## 4.3.5  Principle #5: Manufacturing Systems Transform Inputs into Outputs

The fifth IMP principle asserts that a manufacturing system is a transformation process that transforms inputs into outputs, as illustrated in Figure 4-10. It gets its main ideas from the sociotechnical design approach and the SADT methodology discussed earlier. The principle proposes that the essential aim of the system can be described in terms of having to transform inputs into outputs through the use of various activities which utilize various resources (people, machines, techniques) with the minimum possible error. The process is seen as a continuous activity of transforming input into output. In an NC machine cell, for example, the inputs of raw materials and production schedule are transformed into outputs (goods and status information) through the use of various resources (NC Machine, Operator, Floor Space).

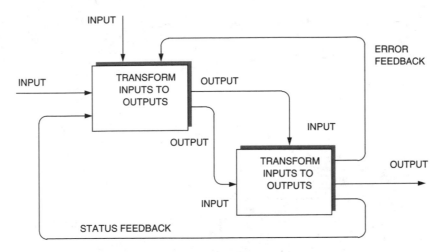

**FIGURE 4-10.** Open systems with feedback.

This principle further proposes that the system can be self-regulating to some degree through the use of various types of feedback. The concept of feedback acknowledges that the transformation process can be changed or altered as a result of knowledge gained during a previous transformation process. Feedback can be of two types—status feedback, and error feedback. Status feedback in the context of a manufacturing activity can be used to confirm the status of the system. For example, a schedule requires that ten parts be produced in a given hour. Status feedback after the hour will confirm that the order is complete.

Error feedback is used to relay information about mismatches between desired output and actual output. Again in the context of manufacturing and using the example given above, error status on the expected manufacture of ten parts could give an actual production of perhaps eight, with two delayed due to material shortages. The idea of feedback describes work systems as being capable of self-regulation without excessive supervision.

In summary, this principle encourages the opinion that the analysis of a target system no matter how simple must include the analysis of other systems and that this analysis can be carried out most effectively when the system is described in terms of its activities, flows, and feedback.

## 4.3.6 Principle #6: Manufacturing Systems are Open

The sixth IMP principle contends that manufacturing systems are open. The concept of open systems recognizes that the systems under investigation are open to the influences of external systems, which can be said to exist in two types of environment—transactional and contextual.[10] The relationship between these two types of external systems and the target system in question is illustrated in Figure 4-11.

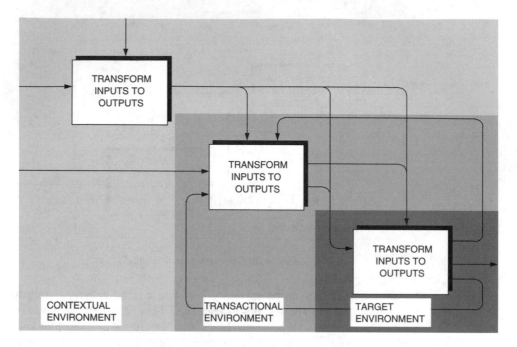

**FIGURE 4-11.** Contextual and transactional environments.

*Transactional Environment:* This represents those systems or activities that have an influence on and can be influenced by the systems being investigated. For example, in the NC machine cell the activities of production scheduling influence the cells operation. The cell in turn can, in certain circumstances, influence the activities of production scheduling as in the case of machine breakdown. This influence can take two forms as intimated earlier in the types of feedback. Error feedback can be a controlling influence on other systems and can directly control the activities of these systems in the contextual environment. Status feedback can form a simple influence on other systems which may be used by these systems only if deemed necessary. In general, the systems in the transactional environment form an essential part of the analysis of any system, although at a more abstract level.

*Contextual Environment:* This environment is characterized by activities that affect the system being analyzed but are in no way influenced by it. The systems are beyond the sphere of influence of the target system. For example, in the case of the NC machine cell, systems such as standards authorities for the design of the cell, customers for the creation of orders, and government regulations for the imposition of safety regulations are beyond the influence of the NC machine system.

### 4.3.7 Principle #7: Manufacturing Systems are Learning Organizations

The final IMP principle contends that manufacturing systems are learning organizations. In the discussion on general systems theory in Chapter 1 this concept was dealt with through the term entropy. Entropy as used here describes a natural tendency for all systems to run out of control. It is perhaps best imagined in the context of decay. A manufacturing system that is not constantly changing both physically (e.g., replacement of machines) and managerially (e.g., policies, strategies, and even personnel) will inevitably develop disorder and begin to decay.

Manufacturing systems avoid the effects of entropy through productive changes in the manufacturing environment. One area that allows change to be fostered and encouraged is the Learning Organization introduced earlier. A learning organization as previously stated is one that *facilitates the learning of all its members and continuously transforms itself.* The issues surrounding the learning environment essentially deal with avoiding the effects of entropy. These issues not only include traditional learning systems such as training but also the encouragement of individual self-development, diffusion of knowledge into the manufacturing environment, making use of the inherent knowledge available at all levels, and extending the learning culture into the customer and supplier environments. Harrison cites an appropriate statement by a businessman in this respect "Our business is learning and we sell the by-products of that learning."[81] This of course is an idealistic statement on a system clearly molded by a learning organization strategy. In reality, a business strategy will be a mixture of the concepts of the learning organization and the more traditional goals of market sector penetration, customer satisfaction, and so on.

The concept of life cycle is closely associated with the effects of entropy. Life cycle deals essentially with the life of a product and individual manufacturing systems (Figure 4-12). The life cycle of these systems may be defined as going through four stages: introduction, growth, maturity, and decline. These stages may be said to occur as a direct result of entropy, and a learning company learns to adapt to these effects and implement further product and individual manufacturing systems to replace decaying ones.

## 4.4   CREATING THE IMP PROCEDURE

Many concepts have been raised in preceding sections; these were divided into two areas. The first area dealt with ideas for change emanating from various well-known writings. The second are the so-called principles of IMP, which can best be described as a synthesis of many existing and emerging methodologies and theory.

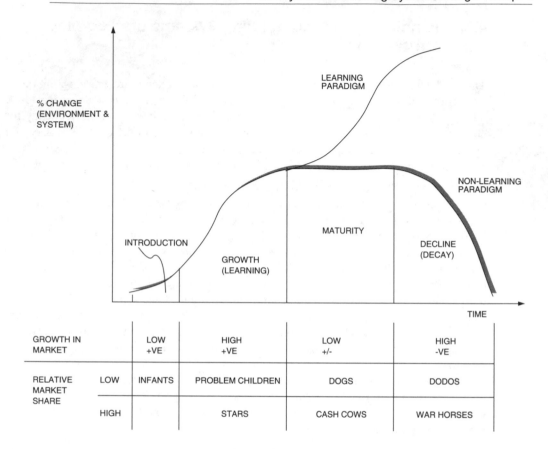

**FIGURE 4-12.**  Learning and life cycle.

The new IMP procedure must reflect as many of these ideas as possible. It must at the same time provide an unambiguous, easy-to-use procedure which designers can follow. In the discussion on the IMP procedure that follows, each of the principles outlined earlier can be either found in, or be derived from, the IMP procedure. Some indication of the way in which many of these ideas are used in the procedure is indicated in Figure 4-13. Many of the ideas for change are also evident in the IMP procedure. These other ideas and concepts are partially derived not only from the use of the IDEFo methodology, but also from the way in which the IDEFo model itself has been created.

## 4.5    CONCLUSIONS

In this chapter we have presented the theoretical foundation of IMP. We now understand that IMP has evolved out of a need to synthesize a number of concepts in the ar-

| Ideas and Principles | Some examples of how the IMP procedure uses this idea or principle |
|---|---|
| **IMP PRINCIPLES** | |
| 1. Goal Seeking | Goals are hierarchically decomposed down through the procedure hierarchy |
| 2. Holistic | Procedure encourages analysis of all major areas |
| 3. Hierarchical | Procedure promotes use of structured analysis techniques |
| 4. Technical / Social | Procedures adopts sociotechnical analysis techniques |
| 5. Inputs / Outputs | Procedure promotes use of structured analysis techniques |
| 6. Open Systems | Context stage defines primary flows across boundary of new system |
| 7. Learning Systems | Procedure is continuous, uses feedback from shop floor as well as from major business activities |
| **Ideas for Change** | |
| 1. Management Responsibilities | Controls inputs and information outputs for the development activity |
| 2. Policy for Change | Change policy used as controlling input to development activities |
| 3. Open Ended Design Process | Use of feedback and activity boxes indicate continuous change process |
| 4. Project Leader | Project leader a key resource for project integration and specialist selection |
| 5. Project Organization | Project organization illustrated clearly in the model |
| 6. Conflict Resolution | IMP promotes training in the area of group dynamics and social integration |
| 7. Self Design | Users clearly indicated as part of the project group |
| 8. Resistance to Change | Promotion of users in the design group |
| 9. Structured Methodologies | Use of key structured methodologies indicated at appropriate stages |
| 10. Abstraction | Project development stages promote analysis of various system views |
| 11. Data Standards | Key standards indicated as information inputs at appropriate stages |
| 12. Equipment Standards | Key standards indicated as information inputs at appropriate stages |
| 13. Group Technology | Indicated a key resource at appropriate stage in design process |

**FIGURE 4-13.** Ideas for change and the IMP procedure.

eas of social and technical integration. Of particular significance in the creation of IMP are the concepts of open systems, structured analysis, learning organization, and socio-technical design. The theoretical part of IMP allows the designer to view manufacturing systems in a holistic way. Figure 4-14 illustrates how the ideas and principles of IMP are used in the creation of the IMP procedure. Also shown is how the IMP procedure together with various social and technical planning tools as well as the systems design group are used as key resources in the creation of new integrated systems. These new systems are ultimately used as a resource in the creation of products in a manufacturing system. Chapter 5 details the IMP procedure and in particular the IMP road map or model of IMP.

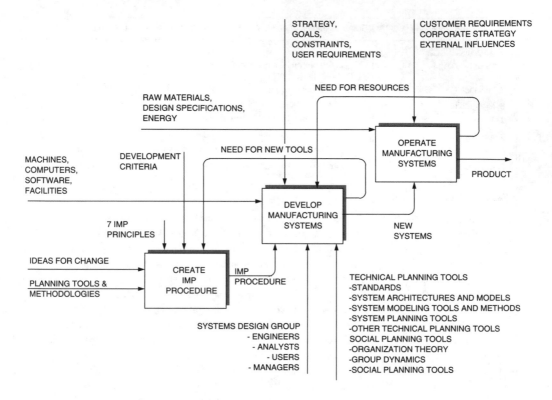

**FIGURE 4-14.** Creating the IMP procedure.

# Integrated Manufacturing Systems Design Road Map (IMP)

**INTRODUCTION**

The IMP procedure presents us with the implementable part of the IMP methodology. It is a procedure in that it has been designed to augment the manufacturing systems design process by guiding the design group in the development and implementation of manufacturing projects on the shop floor. The procedure consists primarily of the IMP model that has been created using the IDEFo methodology. The purpose of the IMP model is to act as a road map for the stepwise introduction of new systems into the manufacturing environment. In essence, the IMP model shows us an idealized design procedure which takes cognizance of the *ideas for change* and *IMP principles* presented earlier. The IMP model is supported with associated text and diagrams which are designed to augment the learning process for the manufacturing systems design group and to facilitate the planning and development of new projects.

The IMP procedure fulfills a number of functions. First, it imparts knowledge or learning about manufacturing systems design to the design group. This results from the need to create a more learning-centered environment. It also recognizes that large manufacturing projects are infrequently implemented, thus hindering the development of the learning process. Second, the IMP procedure presents a systematic approach to the development and implementation of complex projects. This is done primarily through the indication of key stages in project development, from identification of key projects, through their development and on to implementation. Third, the IMP procedure provides us with the unambiguous presentation of key project information. This is indicated through activity inputs, outputs, resources, and controls. A major part of this presentation of key project information is the indication of various tools to be used at various stages. Finally, the fourth function that the IMP procedure fulfills is holistic systems design. Through the use of various checklists and tables which are used in support of the IMP model, the IMP procedure ensures that no major issues "fall between the cracks." The procedural part of the IMP methodology is illustrated in Figure 5-1.

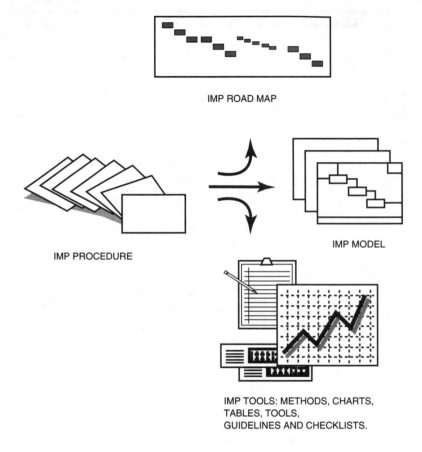

FIGURE 5-1.   Procedural Components of the IMP Methodology

## 5.1.1 The IMP Model

As mentioned earlier, the IMP procedure is a model of an idealized project development process, which together with various charts, tables, lists, and glossaries acts as a *road map* for the manufacturing manager, project manager, and project design group in the development of the project. As a road map it indicates various inputs and outputs to the various project development activities. The IMP model is identified as having five general stages in the project development cycle. These are illustrated in Figure 5-2.

The first three stages—context, systems, and projects development—can be said to be within the activity domain of the manufacturing manager or strategist. These activities principally involve translating the business strategy into a set of time-phased projects to be implemented by the manufacturing systems design group. The next two stages—project integration and project stages—cover the activities to be carried out by

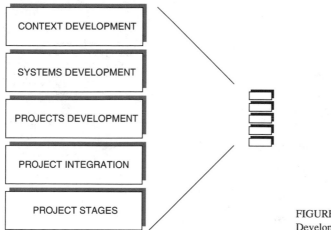

FIGURE 5-2.    Stages in Project
Development

the design group in taking a general project specification and translating it into a physical manufacturing system on the shop floor. These two activities also represent project development activities such as modeling information flows or modeling facilities layouts. The IMP model maps these activities with respect to the required inputs/outputs, controls, and resources. The IMP model identifies thirty four specific activities which need to be carried out. Each of these 34 activities and the relationships between them are shown in summary form in Figure 5-3. Each of these activities will now be discussed under section headings of the five key stages listed earlier: Context Development, Systems Development, Projects Development, Project Integration, and Project Stages.

## 5.2    CONTEXT DEVELOPMENT

The first general stage in the IMP procedure is context development for the activity of developing manufacturing systems. The primary concern at this stage is to develop the context for manufacturing within which manufacturing change may come about. The primary factor in the context of the development activity is the business strategy that outlines company goals, but other factors exist. Context development is not therefore about developing the context for a project, but rather for the system as a whole. Figure 5-4 shows the two environments that generate various inputs to the activity entitled *develop manufacturing systems*. These two environments are the transactional and contextual environments, and each produces inputs that affect the development activities. Two major principles are used in the creation of this stage of the IMP model.

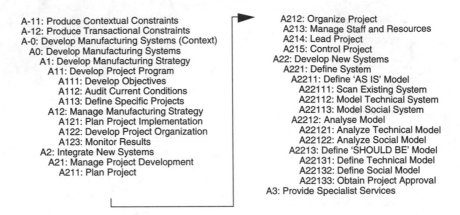

A-11: Produce Contextual Constraints
A-12: Produce Transactional Constraints
A-0: Develop Manufacturing Systems (Context)
A0: Develop Manufacturing Systems
  A1: Develop Manufacturing Strategy
    A11: Develop Project Program
      A111: Develop Objectives
      A112: Audit Current Conditions
      A113: Define Specific Projects
    A12: Manage Manufacturing Strategy
      A121: Plan Project Implementation
      A122: Develop Project Organization
      A123: Monitor Results
  A2: Integrate New Systems
    A21: Manage Project Development
      A211: Plan Project

A212: Organize Project
A213: Manage Staff and Resources
A214: Lead Project
A215: Control Project
A22: Develop New Systems
  A221: Define System
    A2211: Define 'AS IS' Model
      A22111: Scan Existing System
      A22112: Model Technical System
      A22113: Model Social System
    A2212: Analyse Model
      A22121: Analyze Technical Model
      A22122: Analyze Social Model
    A2213: Define 'SHOULD BE' Model
      A22131: Define Technical Model
      A22132: Define Social Model
      A22133: Obtain Project Approval
A3: Provide Specialist Services

FIGURE 5-3.   Node Diagram for the IMP Model

The first of these principles is that systems are open and therefore subject to the traits of open systems theory. These traits state that open systems are: Holistic, Hierarchical, Transforming, Energy Creating/Consuming, Entropic, and Cybernetic. This view of manufacturing systems design was discussed in Section 1.3.2 and later used for the development of the IMP theory discussed in Section 4.3. Figure 5-4 is, in effect, a systemic view of the activity being examined. The second principle is that, within the environment of manufacturing systems design, two types of environment may be identified—the transactional and the contextual. The concept of these two environments comes from the sociotechnical design technique discussed in Section 3.5.1.

A number of other principles and ideas are apparent from Figure 5-4 including: transformation of inputs into outputs, hierarchical decomposition, hierarchy, feedback, holism, and distinction between controls and inputs. These principles and ideas also apply to all of the other diagrams or levels of IMP to be described throughout this chapter.

## Produce Contextual Inputs (A-11)

The *produce contextual inputs* activity represents the contextual environment for the activity *develop manufacturing systems*. The contextual environment represents external groups or activities that are beyond the sphere of influence of the core activity. For example, in the case of developing a production system, external influences will include: customer order variances, standards development, systems development by competitors, new process/product developments, new governmental directives, vendor availability and reliability, and techniques and methods for systems design. These externally created inputs can have a major impact on the development activity. However, the core activity (develop manufacturing systems) can do little to influence these inputs. Recognizing the existence and influences of these inputs is important to the systems development cycle and for preparing plans for project implementation.

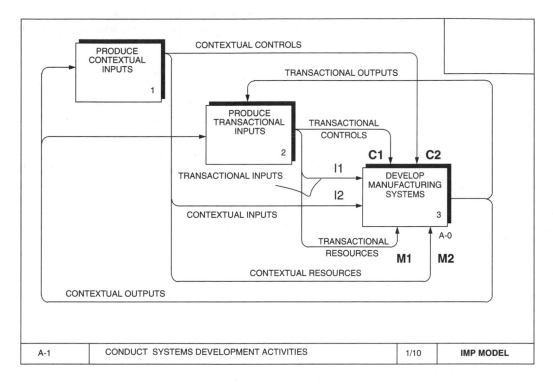

| A-1 | CONDUCT SYSTEMS DEVELOPMENT ACTIVITIES | 1/10 | **IMP MODEL** |

FIGURE 5-4.   Conduct Systems Development Activities

Many contextual inputs to the development activity have been discussed and documented in Chapters 2 and 3. Planning systems such as IDEFo, sociotechnical design, and GRAI are all important available inputs to the development activity. As a first step, context development facilitates the recognition and understanding of these planning systems. Later, the design group may decide to use many of the systems in implementing manufacturing systems projects.

## Produce Transactional Variants (A-12)

The second environment that affects the development activity is the transactional environment, which produces various inputs to the development activity. In turn the development activity can produce outputs that can change the behavior of the transactional environment. Two primary inputs from the transactional environment for manufacturing systems design are the business strategy and user requirements.

The business strategy will be created by a top management activity in the transactional environment. Outputs from the development activity such as status reports, the manufacturing strategy, and results of various audits can help change the business strategy. The new business strategy can then be seen to synthesize the knowledge and information originating from the lower management levels with business objectives. Note that the output from the development activity cannot control the transactional environment. The output is mainly used by activities within the environment during their own transformation process.

### Develop Manufacturing Systems (A-0)

The final activity in Figure 5-4 represents the core activity (develop manufacturing systems). In short, this activity may be said to transform contextual and transactional inputs (inputs, resources, and controls) into outputs. Isolation of the development activity from its two environments produces the diagram illustrated in Figure 5-5, which shows some of the main inputs and outputs to the development activity. This view of the development activity allows us to isolate the target activity and map the flows crossing its boundary. This activity is discussed in more detail in Section 5.3.

***Information Requirements:*** The main information requirements for the Context stage of IMP are categorized as follows:

1.  Description of the company, its products, markets and major facilities

2.  Description of the business mission and major performance parameters

3.  Description of business goals and strategies

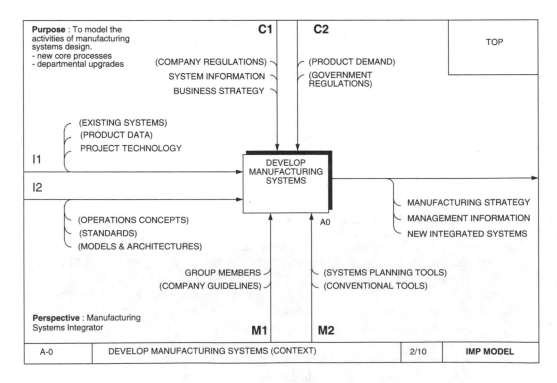

FIGURE 5-5.   Develop Manufacturing Systems (Context)

The information generated in each of these three categories produces a contextual view of the business for the design group. In the context of the operating system being designed additional information requirements are as follows:

**4.** Description of upstream and downstream activities

**5.** Identification of major variables (inputs and outputs)

The net effect of the information gathered in each of these five categories is to produce a focus and context for the system being designed.

*Architecture Development:* Many large companies, and even some medium-sized ones, have identified an approach to systems development based on the concepts of architectural development. This approach, sometimes called the Architectural Development Approach, can be said to be loosely based on the techniques used in many of the collaborative research projects discussed in Section 2.3.2. This approach, when used in the context development stage of IMP, yields architectures for the manufacturing strategist that can be used as role models and frameworks within which particular projects can be implemented. The architecture development approach consists of four steps: (1) Model, a step involving the creation of models (information and other) of the ideal environment, (2) Architecture, a step involving the creation of accompanying rules and guidelines for managing the development of systems design, (3) Design, incorporating the development of the many individual projects that adhere to the architecture, and (4) Realization, a step involving the determination of physical activities for implementing the architecture through its many projects.

The architectural development approach in a sense creates a very formal context for future systems design. It provides a generalized model of the ideal system, portraying the essential structure, functions, content, and boundaries. The architecture provides general rules on the transformation process over the long term. The essential result in carrying out the architectural development approach is to give a generalized view of the target environment being designed and to inform designers of long term-yet generalized goals. A common example of a long-term architecture is that associated with the development of a computer network. Guidelines are required for the possible future layout of the network, its long-term requirements, the type of computer hardware best suited for the job and scope for future expansion. A less common example is an architecture that provides models and guidelines on the flow of information, its data structure, and key decision points.

## 5.3  SYSTEMS DEVELOPMENT

The second general stage of the IMP procedure is systems development. This stage primarily represents ideas on the organizational structure and various flows in the manufacturing systems design function.

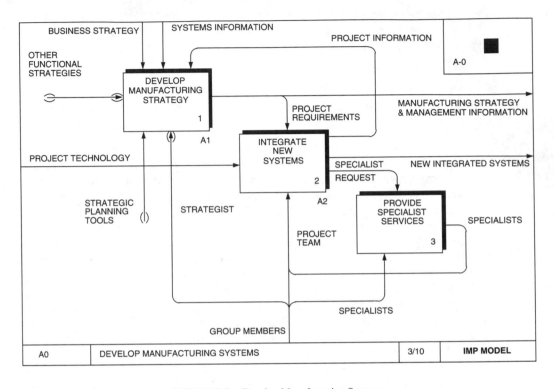

FIGURE 5-6.    Develop Manufacturing Systems

Figure 5-6 shows a decomposition of the activity *develop manufacturing systems*. In this figure the development activity is divided into three separate activities to reflect a new type of social structure within manufacturing systems design. The social structure reflects two main ideas, the first idea of which comes from the research and surveys of the SME into the roles of manufacturing personnel.[2] This survey, which was discussed in Section 1.2, identified three distinct roles—the project strategist, the project integrator, and the project specialist. The strategist is seen as someone with good breadth and little depth of knowledge, who is able to create manufacturing strategies. The specialist has good depth of knowledge but little breadth and is suited to special tasks. Finally, the integrator is someone with both good breadth and good depth of knowledge. His or her role is to integrate various aspects of the project in particular social aspects. These three roles are identified in Figure 5-6 as activities 1, 2, and 3, respectively.

Another concept used in the creation of the model at this level is that project teams are to be organized around a *heavy weight project manager* identified by Hayes and discussed in Section 3.3. In the model various control flows are illustrated that indicate the type of control Hayes sees as necessary for proper project organization. For example, specialists are requested from various functional areas by the project integrator. These specialists then form part of the project integration activity and report directly to the project manager on project-related issues. Many other ideas included in this stage of the IMP model are: (1) involving users in the project group, (2) use of *group* dynamics, (3) decomposition of business strategy into manufacturing strategy, (4) project requirements, (5) feedback between major activities, and (6) the use of various planning tools.

### Develop Manufacturing Strategy (A1)

This activity primarily represents the work of a manufacturing strategist. The main activities carried out are to translate the business strategy into a manufacturing strategy, which may then be translated into a facilities strategy and eventually into manufacturing projects. The decisions that are made must take cognizance of systems information, which includes user or operations personnel, project requests, and the equipment and systems that are already available. Project budgets, existing project status, and project performance requirements are also important data flows into this activity. Various tools that have been mentioned earlier can be used as resources in this activity. These include strategic planning tools as well as traditional mathematical methods for prioritizing project preferences and making sure they match the manufacturing strategy. This activity is discussed in more detail in Section 5.4.

### Integrate New Systems (A2)

The second activity represents that of a project group which has responsibility for integrating the new project under the budgetary and performance constraints of the manufacturing strategist. The project technology in this figure represents the hardware, software, and methods that are purchased by the group and translated into an operating system, before handing the new system over to the operations personnel. The project team or group itself is made up of users, specialists, and a project manager. This activity is discussed further in Section 5.5.

### Provide Specialist Services (A3)

The final activity indicated in Figure 5-6 represents the various specialist groups that exist in a company. These specialist groups play an important role in the functions of management information systems, maintenance, operations planning and control, human resources, quality, marketing, and accounting. Each functional group is normally a separate department which develops various strengths and levels of expertise. The participation of each functional group in the project group provides a wide cross section of expertise. These specialists can be requested to participate in the project group as full members. Alternatively, they may be drafted into the project during critical stages of project implementation. Various specialist equipment including programming languages, CAD systems, materials laboratories, and testing laboratories will be used by these functional groups.

*Information Requirements:* The main information requirements for the Systems Development stage of IMP can be described as follows:

1. Description of systems design mission and major performance parameters
2. Formation of project group and statement on group organization

The generation of this information is important in setting the initial standards used in the design group. In the context of the operating system being designed, additional information requirements are:

3. Description of system, its products, customers, and main facilities

4. Description of systems mission and major performance parameters

5. Description of systems goals and requirements

6. Description of main conflicts and interactions with other systems

The information generated in this stage of IMP gives a broad image of the project group's reason for being and a description of the environment being designed.

## 5.4    PROJECTS DEVELOPMENT

All of the activities in this third stage of the IMP procedure may be said to contribute to the development of a manufacturing strategy and in particular a set of projects to be implemented by various design groups.

The development of the manufacturing strategy is described by a two-stage process illustrated by the two activities shown in Figure 5-7. The first stage transforms the business strategy as well as systems information into a facilities plan. This plan outlines the various projects to be developed in order to meet the business strategy and user requirements. The second stage transforms the facilities plan along with budget constraints into a set of time-phased projects. It also begins the process of preparing a general project group organization and identifying the project leader. The end result of both of these activities is a set of projects to be developed and implemented according to predefined performance parameters and predetermined time schedule.

The main idea built into the model at this stage ensures that projects, no matter how small, reflect the constraints of the business strategy and systems information (in particular, user requirements). This concept was discussed in Section 1.4. The use of structured decomposition in the IMP model ensures that the manufacturing strategy *flow* is clearly decomposed into smaller more detailed flows. The facilities plan, for example, is the product of the transformation of the business strategy. Later the facilities plan and many other details are joined together to represent the manufacturing strategy. Other ideas in the model at this stage include: (1) the use of planning tools, (2) feedback to higher management, and (3) continuous monitoring of process information through *systems information* flow.

### 5.4.1 Develop Project Program (A11)

The *develop projects* activity derives its main controlling input from the business strategy. This is done preferably through exchanges in documentation but also in other unstructured ways, such as through meetings and discussions. The business strategy indicates a broad strategy for changes to the manufacturing system in line with any mar-

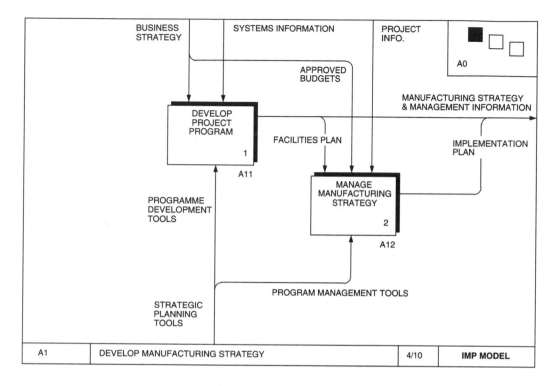

FIGURE 5-7.   Develop Manufacturing Strategy

keting and financial strategies. Another important control on this activity is system information, which in part requests changes to the manufacturing environment. While the strategic planning is essentially a top-down decomposition, the ideas for change will tend to come from the bottom. The request for changes to the manufacturing strategy is a critical bottom-up feedback in this process. It not only provides ideas and concepts for change but also involves the very people for which changes in the manufacturing system have the most effect.

The output from the *develop projects* activity is the manufacturing strategy that, like the business strategy, is partially reconstructed in documents such as facilities plans but mainly through the informal interaction between the manufacturing strategist, the project teams, and other management personnel. Figure 5-8 shows the decomposition of the "develop projects" activity and illustrates three new activities.

## Develop Objectives (A111)

The *develop objectives* activity is concerned with setting realistic objectives that can be measured and monitored. These objectives primarily act as a guide for the development of a number of projects. Objectives are statements about the project and should be translatable into real savings for the manufacturing company. The ultimate aim of

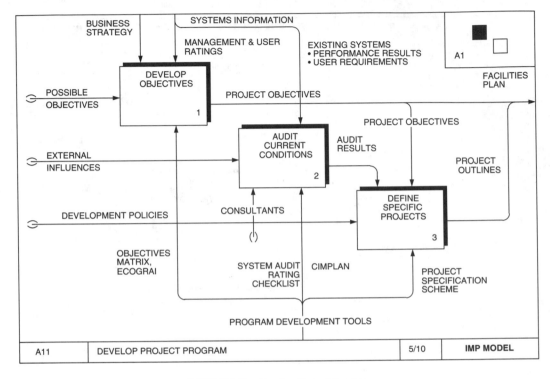

FIGURE 5-8.   Develop Project Program

achieving objectives is to increase the efficiency of the manufacturing system. There are three general classifications of objective: (1) capacity objective, (2) performance objective, and (3) intangible objective.

**1.**   Capacity Objectives relate primarily to increases or decreases in machine and labor capacity. The objective of a project might be to either increase or decrease the capacity of a system. Costs are offset by decreased production costs (decrease in capacity) or increased contribution to profit (increase in capacity).

**2.**   Performance Objectives are objectives that can often be coupled with capacity objectives and require some improvement in certain performance parameters to justify the costs incurred in the project. There can be numerous performance objectives.

**3.**   Intangible Objectives are productivity objectives that cannot easily be translated into cost savings. Objectives such as Improved Management Information or Improved Operator Responsibilities cannot often be easily quantified. Intangible benefits are seldom used solely to justify a project and are often coupled with the other two types.

Some objectives can overlap with others. On the other hand, some objectives can be in contradiction with others. The important thing for the manufacturing strategist is that the right objectives are selected for a project and that achievement of these objectives is in keeping with the overall objectives of the company.

*Rating Objectives:* For the reasons explained above, not all objectives will carry the same weight. In addition, they may receive differing priorities through interpretation of the business strategy. Therefore some kind of rating scheme is required for prioritizing and classifying the objectives. A rating scheme is one way in which the manufacturing strategist may involve other managers in creating the manufacturing strategy. By asking other managers to add to the list of objectives and then talking through their ratings of each objective, the strategist can enhance his or her view of the business strategy and soften resistance to a particular project. Rating schemes can vary from company to company. One scheme that is useful and that also facilitates the measurement of the objectives is the Objectives Matrix.[82] A useful approach to the stepwise translation of the business strategy into project objectives is also presented in a tool from GRAI called ECOGRAI.

## Audit Current Conditions (A112)

*Audit current conditions* is the second activity in the development of the project program. The main inputs are the two controlling inputs—business strategy and existing systems. In order to audit an existing system, it is necessary to initially form a perspective on what exactly should be audited. The selection of what to audit and the degree of detail will be determined first by the information contained in the business strategy and second by the experience of the strategist. Auditing involves gathering various information about the system. Equally important, it involves gathering details of requirements from users on the shop floor. Most of this information can be obtained from good reporting systems. However, there will be times when special information is required.

In the *audit current conditions* activity of Figure 5-8, two particular resources are indicated. The first is the use of outside consultants to carry out much of the auditing function. Two types of consultants are available. The first is the "expert" consultant, who arrives at the company with a so-called *bag of tricks*. After extensive interviewing and presentation, he leaves the company with a solution to the problem but with little or no knowledge on how it was solved.

An alternative type of consultant is the *facilitator*, who is a consultant with a certain degree of know-how, but equally important has a willingness to learn about the company's systems, and facilitate their change. The facilitator strives to teach personnel within the company how to carry out the changes themselves, what tricks are available to them, and to a lesser degree than the expert consultant he carries out some of the changes himself. The facilitator leaves the company not only with deliverables in terms of reports, hours worked, etc., but most importantly he imparts to the company personnel the knowledge and confidence to carry out the remaining work themselves. Current trends suggest that using this latter type of consultant is the most beneficial in the long run for the development of a company's systems (see Section 4.2.3 for key activities carried out by a facilitator).

The second resource indicated in Figure 5-8 is an *audit rating scheme*, which is provided by a number of tools, some of which were discussed earlier. One such tool is CIMPLAN as discussed in Section 2.5.1. A second tool useful for auditing and in particular for creating a framework under which more detailed audits can take place is the GRAI-Grids technique discussed in Section 2.4.1. The *manufacturing audit checklist process* as is described by Skinner details a large number of audit parameters mainly aimed at a general nonstrategy related audit.[1]

### Define Specific Projects (A113)

Identification and definition of specific projects is the third activity in the development of a projects program. This activity is critical to the whole development cycle of the project. It is at this point that specific projects are identified as being important to the strategic goals of the business. The number of available alternatives will vary considerably among functions, departments, and companies. It will also vary according to the personnel carrying out the definition procedure. For this stage to be effective certain *development policies* will be used by the manufacturing strategist. These development policies indicate issues such as:

**1.** Technology purchasing policies: How long do we wait before replacing machines? What level of automation is appropriate for our plant? What types of technology does our plant need? What is our technology strategy for the future?

**2.** Make-versus-buy policies: Do we tolerate major subcontracting? Is subcontracting an alternative?

**3.** Automation versus human policies: Do we automate all processes? Where do people fit into our manufacturing strategies? What is our attitude towards people and technology?

A project specification scheme from CIMPLAN (discussed in Section 2.5.1) is a useful resource for this stage of projects development. The outcome of the definition stage will be a set of projects and their outlines. The outline of the project will typically include the following elements:

**1.** Purpose and objectives

**2.** Scope and approach

**3.** Deliverables and results

**4.** Resources and time frame

**5.** Relationships with other projects

These project outlines are the basis for individual project directives for the project group yet to be selected.

***Development Funnel:*** A useful way of conceptualizing the manner in which project ideas are generated and selected by the manufacturing strategist under various constraints is illustrated in Figure 5-9.

This figure shows many suggested projects placed at the mouth of the funnel. Projects that eventually move into the main flow are those that satisfy the constraints set by manufacturing strategy, financial considerations, and resource availability. The rate at which projects are eventually implemented is also constrained. A fourth constraint shown in Figure 5-9 is the requirements of a manufacturing architecture. In effect, projects that flow into the funnel must adapt to the requirements of the architecture so that they contribute to an integrated manufacturing environment.

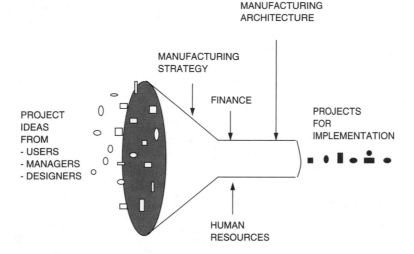

FIGURE 5-9. Development Funnel

## 5.4.2 Manage Manufacturing Strategy (A12)

The second stage of the manufacturing strategy development process, first illustrated in Figure 5-7, is management of the manufacturing strategy. This stage consists of three additional activities as illustrated in Figure 5-10. The three activities convert the facilities plan created in the first stage of manufacturing strategy development into an implementation plan, which reflects the specific project outlines, their duration and timing, the resources allocated (people and money), and the project organization responsible for implementing each project.

An important concept, which is not apparent from the IMP model, is that activities are not necessarily sequential, nor are they time dependent on each other. For example, the *monitor results* activity in Figure 5-10 can be an ongoing process. The actions of the activity may only change with new information received.

### Plan Project Implementation (A121)

Planning the orderly release of projects to the development and implementation stages is the fourth stage in the development of the manufacturing strategy. The way in which this plan is created will reflect the need to balance a number of issues, which are:

1. The need to achieve objectives that address the business strategy
2. The priority given to each objective
3. Budgets being made available to address each general area
4. Staff skills and staff availability
5. User requirements and user receptivity to changes to the working environment

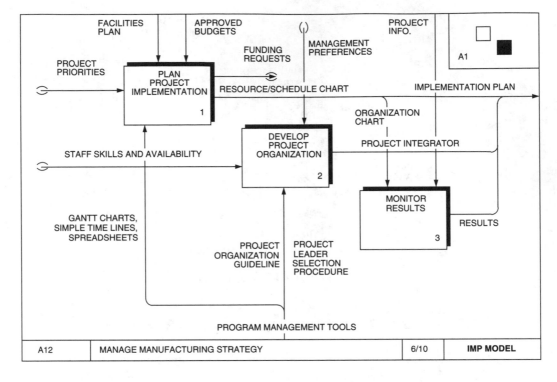

FIGURE 5-10.   Manage Manufacturing Strategy

The skills developed by the manufacturing strategist will be the most important resources during this stage. However a number of pointers are appropriate:

1.   Crawl, walk, run—tackle projects that have a good change of success

2.   Use rough-sequencing tools for project scheduling

3.   Generate many small projects rather than one *showcase* project

4.   Overlap individual project introduction

One type of resource is the simple time line. Gantt and PERT charts are also useful.
Gantt and PERT, however, require much maintenance and initial coding. At this stage
of the planning process, these techniques are more cumbersome then the simpler time
line.

***Manufacturing Development Plan:*** Often when a number of projects have been
identified they are presented to top management for outline approval and budgeting. It is
often an iterative process with top management mainly deciding how much to invest
within what time period. The plan, once approved by management, can be implemented,
updated, and modified over the period in question (typically up to three years). The con-
tents of a typical plan are as follows:

1. Executive Summary including corporate mission, quality, health, safety, and people statement.
2. Current Status including a review of capacity issues, social issues, and performance
3. Future Requirements and how they match with business mission
4. Main Proposal or overall Architecture of new system for coming three years
5. Projects including general description, objectives, advantages, risks, and costs
6. Timetable and Implementation Plan
7. Resource allocation and requirements (group members)
8. Financial Analysis and Break Even
9. Sensitivity Analysis
10. Recommendations

Appendices can include: Sales Forecasts, Cost Studies by Project, Details–Sensitivity Analysis, Layouts and Plans, Details–Resource Requirements, Key Process Specifications, Construction Costs, Key Suppliers, Competitor Information, and Miscellaneous.

### Develop Project Organization (A122)

The next stage is the development of organization for the projects being executed in the near term. One of the first steps will be the selection of the project manager. The characteristics of the project manager and the best type of project organization were discussed in Section 3.4. The selection of the most suitable project manager can be a difficult task and as discussed earlier is one of the key links in proper project implementation. Some of the qualities desirable in a project manager that should help in his or her selection, are:

1. Middle/higher management experience
2. Intimate working experience of the company and company strategies
3. Keen interest and enthusiasm for systems design
4. Ability to lead and motivate project group members
5. Integrity and attention to detail

A formal approach to the selection of a project manager is available entitled Project Leader Selection Procedure developed by Margerison et al.[83]

### Monitor Results (A123)

Monitoring of project information and the generation of results for manufacturing management is the final activity in the management of the manufacturing strategy. Some interesting aspects of this monitoring process are the typical phases that projects undergo during their life cycle. These phases are often identified as Optimism, Struggle, Disaster, Regrouping, Compromise, and Completion.

Projects usually begin with great optimism on the part of the project manager and project group. When difficulties arise, the project manager and project group spend most of their time fighting these difficulties and maintaining enthusiasm. Disaster can arrive when the project seems to have reached an insurmountable problem. The project is momentarily halted until people can determine what the problem is and how it affects the rest of the project. A regrouping phase acknowledges that some problems have to be ignored and that the project's success will have to be measured by other features. This compromise between what was expected and what is achievable ultimately leads to project completion. Awareness of these phases can help avert potential problems within the project organization and allow management to *learn* across development projects in order to manage situations that have become described as *disasters*.

## 5.5   PROJECT INTEGRATION

Project integration is the fourth general stage in the development of manufacturing projects. This stage emphasizes the activities of the project manager or project integrator. Figure 5-11 shows the hierarchical decomposition of the activity *integrate new systems* first introduced in Figure 5-6. Two activities are shown in the decomposition— *manage project development* and *develop new systems*.

The main concepts to be seen at this level of the IMP procedure are the separation of project management from the project development stages. This separation allows us to concentrate on the features of each activity in more detail. Many current texts and tools that claim to offer a means or procedure for handling project integration merely address the activities of *manage project development*. Software tools for resource scheduling, task scheduling and so on, are simply project management aids. As we shall see in Section 5.6 many other tools and techniques are available for project development. In most projects these tools are equal to if not more important than project management tools for project integration.

### Manage Project Development (A21)

The activity *manage project development* represents the activities of the project manager whose primary responsibility is to ensure that the project is integrated into its social and technical environments. This integration must take place under the constraints of the project requirements issued by the manufacturing strategist. The way in which the project will become integrated will depend largely on the perspective of the project manager and the way in which the project manager deals with each of the issues involved. For example, by adopting the strategy of using structured methodologies, certain standards, sociotechnical design, and perhaps dynamic workgroups the project manager is making a decision which will ultimately decide the success of the project.

Various tools can be used by the project manager for the execution of the various tasks. Project management software packages are available for project budget, schedule, and resource control. Typically tools such a word processors, graphics processors, chart processors, and spreadsheets will suffice for most activities.

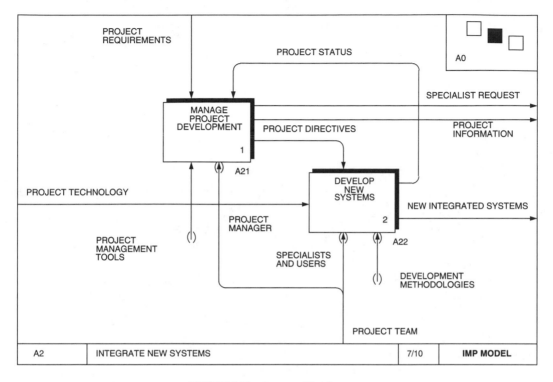

FIGURE 5-11. Integrate New Systems

Looking at the activity *manage project development* in isolation and breaking it into smaller activities gives us the arrangement illustrated in Figure 5-12. In this figure five activities are identified as being a part of project management. This five-stage classification comes from an original classification developed by Koontz.[67] The resources, controls, inputs, and outputs reflect the overall progress of the development process.

*Selecting a Project Group:* The selection of the project group members is primarily based not only on the functional expertise required by the project, but it will also require the selection of people who are able to work in groups. A general guideline to the selection of group members is as follows:

1. Identify and define skill requirements

2. Identify which skills are required at what stages of the project

3. Identify available in-house skills or potential skills

4. Identify training needs to increase in-house skills

5. Identify out-of-house skills requirements (consultants, researchers, employees from other plants)

6. Organize project around core management group—subgroups may also need to be identified

7. Obtain approval and support from key functional managers and commitment to allow *their people* to give loyalty and total commitment to the project group

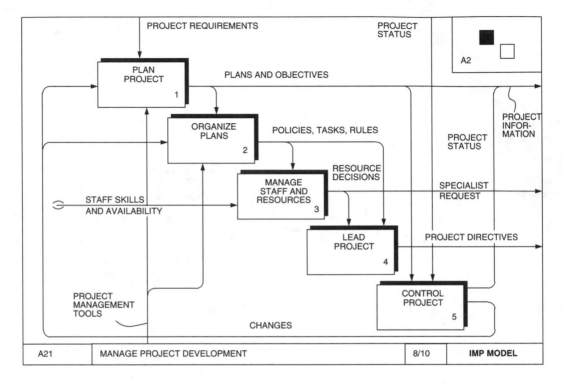

FIGURE 5-12.    Manage Project Development

### Develop New Systems (A22)

The second activity *develop new systems* represents the core of project work associated with project development. This activity is primarily constrained by project directives, which are issued by the project manager based on the initial project requirements and the project status. Project technology (hardware, software, specifications, etc.) are translated into the new system in progressive stages which will be discussed in Section 5.6. As mentioned earlier the resources to be used in this activity may be a combination of structured methodologies and sociotechnical design techniques. Activities in this area are carried out by the project group which as mentioned earlier will consist of specialists and users. As with many of the activities, information flows are in two directions, one forward and the other feedback.

## 5.6    PROJECT STAGES

Project stages is the fifth and final general stage in the IMP procedure. This general stage outlines generic project development stages. The idea of outlining project stages is not to impose a strict structure of the stages to be used in project implementation.

One of its aims is to provide a useful guideline for the systems designer in the creation of his or her own specific stages. Another aim is to promote various concepts developed through the IMP theory. Some of these concepts include the idea of project stage overlapping and the principle of systems consisting of social as well as technical subsystems. Figure 5-13 shows three generic stages for the activity *develop new systems*.

### Define System (A221)

The *define system* activity brings out a number of key ideas and principles for the project development activity in general. One idea is the use of structured methodologies as a primary resource in the modeling of the new system. Another idea concerns the use of social design techniques for augmenting the design of the social and technical subsystems. The *define system* activity is divided into the following three activities: (1) *define existing model*, (2) *analyze model*, and (3) *define future model*. Within each of these stages an approach to modeling and analysis is used which is a synthesis of some of the systems modeling tools (described in Chapter 2) and the social planning tool of sociotechnical design. The latter facilitates the concept of minimum critical specifications to allow the group quick access to key project information. System modeling tools have been selected that complement this approach and structure the systems information to be analyzed. This activity is dealt with in more detail in Section 5.6.1.

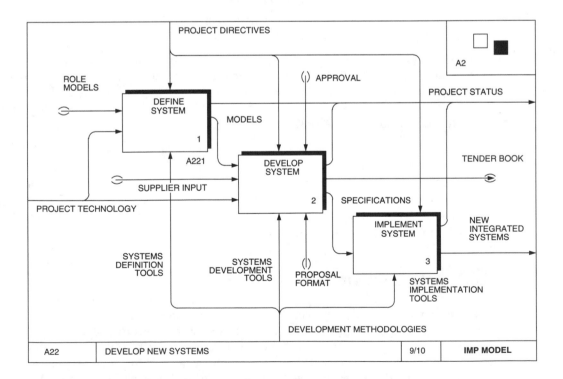

FIGURE 5-13.   Develop New Systems

### Develop System (A222)

Following the definition and approval of the project, the project group enters the development stage of the project. Development is the first of two stages for translating definition models, goals, and available resources into a working system, which must insofar as is possible achieve these project goals under the constraints of the project resources. The development of the system involves developing plans for the creation of new social and technical subsystems. Many of the tasks to be carried out in this activity include: (1) liaison with vendors and systems suppliers, (2) detailing system operation, (3) preparing software specifications, (4) preparing new machine layouts, (5) detailing new job specifications, (6) developing user interface specifications, (7) making adjustments to company organization, carrying out training programs, and creating staged implementation plans.

The development stage is by definition more detailed than the definition stage. The project group will require additional tools to those used during the definition stage for system specification. During this stage it is convenient to define a number of development tasks to be carried out by the design group. These tasks are directly associated with the framework created by the CIM-OSA prenormative standard on systems development. These tasks, which were described in Section 2.3.2, are:

1. Development of the functional subsystem
2. Development of the resource subsystem
3. Development of the information subsystem
4. Development of the organization subsystem

Various tools can be defined as being appropriate for augmenting the development of each of these individual stages. Chapters 2 and 3 contain discussions of some of the more popular tools that may be used in this domain.

### Implement System (A223)

Implement system is the final activity in the development of new systems. Implementation is the process of transforming various project inputs—new machines, new systems, new organizations, new job specifications—into project outputs. These project outputs are collectively called *the new system*. As with the development stage, four primary tasks (listed in the previous paragraph) can be associated with implementation of the new system. However, the tools available for implementation reflect implementation issues.

## 5.6.1 Define System (A221)

The *define system* activity is further broken down into three finer project development stages (as illustrated in Figure 5-14) that promote an architectural development approach. In this approach the existing system is first modeled using the perspective adopted by the project group as a whole and strictly controlled by the project manager. The models created are then analyzed with respect to project goals and state-of-the-art technology. This analysis leads to changes in the models which when coupled with project constraints lead to the development of idealized or future models. These future models form part of the architectural plans for the development of the new system. When these models are approved and released, the project can proceed into the development and implementation stages.

### Define Existing Model (A2211)

The *define existing model* activity represents activities for modeling the social as well as the technical subsystems in the existing system. This process or activity is comprised of three smaller stages: (1) *scan existing system*, *model technical subsystem*, and (3) *model social subsystem*.

*Scan Existing System (A22111):* The first stage of defining an existing system involves an initial scan, which allows the design group to prepare an initial view of the system. The design group identifies influential factors in the environment; specifies the existing system's major inputs and outputs, characterizes the existing system's major historical, social, and physical features; and gathers information on the existing system's aims and objectives and philosophy for management. The scan of the existing system is in effect a context development activity for the operational details of the system being designed. In this respect the application of the context development techniques used in the

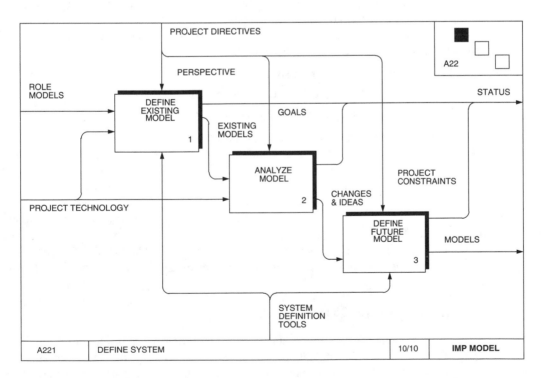

FIGURE 5-14.  Define System

first general stage of IMP are also appropriate here. However, the activity under examination will not be *develop new systems* but rather *operate new system* (see discussion in Section 1.4, *Development and Operation of Manufacturing Systems*).

***Model Technical Subsystem (A22112):*** The second stage in modeling the existing system involves technical modeling of a number of the system's technical subsystems, which can take place at a number of different abstraction levels. Three useful levels discussed in Section 2.4 are used in various methodologies follows:

1.  Modeling the Resource subsystem
2.  Modeling the Information subsystem
3.  Modeling the Data subsystem

A number of modeling methodologies can model each of these stages. Three separate tools may be used. However, it can be more efficient if an integrated methodology comprising a number of tools is used. In this respect the methodology GIM (see Section 2.4.2) is effective in this domain. GIM is also useful for modeling part of the social subsystem.

***Model Social Subsystem (A22113):*** The third stage in modeling the existing system involves social modeling. In this stage the design group models various aspects of the social subsystem. Models will include information on division of labor, decision flow processes, and degree of operator satisfaction. Three separate tasks can be identified which form a synthesis of tasks from sociotechnical design, the human impact infrastructure statement, and Grai. These tasks are as follows:

1.  Modeling the Decision subsystem
2.  Applying Psychological Job Criteria to various jobs
3.  Questioning various aspects of the Human Infrastructure

In this regard three modeling tools are appropriate for each area. GIM, or in particular GRAI, is useful for modeling the decision subsystem. The list of Psychological Job Criteria outlined by Pava is available from the sociotechnical design approach (see Section 4.3.4) to help designers apply various sociotechnical design criteria to the application domain. These criteria help designers evaluate how the new system will impact the social subsystem. Finally, the questionnaire and technique entitled the Human Infrastructure Impact Statement (HIIS) from Majchrzak is available for examining various aspects of the human infrastructure (see Section 3.5.2). Majchrzak developed this questionnaire specifically to address the integration of the technical subsystem into a sociotechnical environment.

### Analyze Model (A2212)

Analyze model is the second stage in the definition of the system and involves analysis of the social and technical models created. An attempt is made to synthesize the information gathered to identify various problems. This stage consists of two separate activities—*analyze technical subsystem* and *analyze social subsystem*.

*Analysis Technical Subsystem (A22121):* Analysis of the technical subsystem involves the analysis of the technical information gathered during technical subsystem modeling. The analysis will involve: (1) looking for variances both internal and external in the system, (2) analyzing various information flows and their effect on the system, (3) anticipating information and physical blockages in the system, (4) understanding the operation of the existing technical system, and (5) preparing changes to the existing system to meet technical goals.

*Analysis Social Subsystem (A22122):* Social subsystem analysis will involve (1) understanding the results of the social models created, (2) identification of real social networks as opposed to formal social networks, (3) contribution of roles to technical variances and (4) formulation of ideas for changing roles and social organization to meet social goals.

In this stage, models are used together with other information for the creation of models for a future system. Other information used with the models will include: supplier technology, role modeling of other factories, expert advice, and company preferences. The output of the analysis stage is a set of results and information which together form the basis of the technical and operational specifications used during the development stage. Analysis information forms the backbone of these specifications and therefore is critical to the success of the project. Two important inputs at this stage are the results of role modeling and supplier knowledge. Role modeling promotes visiting other similar perhaps more advanced factories, trade shows, and conferences and cooperation with various experts and analysts. Supplier knowledge involves working with potential suppliers in analyzing requirements. Suppliers will often have superior knowledge in specialist areas such as machine tool capabilities and materials handling. Most suppliers will be happy to cooperate with companies in this regard in the hope of being allowed to bid for equipment and systems later.

## Define Future Model (A2213)

The preparation of models of the future system, presenting the findings of the definition stage to management, and obtaining final approval for project implementation are the final stages in the definition of the future system. The models of the future system can be created using the same tools and techniques used during the definition of the existing model. A key part of this final stage consists of presentation of findings and obtaining project approval.

Presentation of findings includes results of analysis and proposed future models which are based on knowledge accumulated by the design group. The major criteria for judging the accuracy of the proposed future system is *best match* or *best fit* to the project goals. These project goals must clearly show a link with company goals. The format of the final presentation must clearly cover a number of key issues including: (1) mission and philosophy of company, (2) goals of manufacturing, (3) goals of the project, (4) mission of existing system, (5) definition of system organization and operation, and (6) proposed changes to social and technical subsystems.

Generating imaginative and realistic proposals for change to an existing environment can often be the most difficult stage in a project. Much time and effort will have already been spent on the project by the design group. Management will be eager to test the project group's conclusions before approving a project that will require further major resources, time, and disruption of existing systems.

## 5.6.2 Develop System (A222)

The net outcome of the definition phase are models that can be used in the development of operational and technical specifications. These specifications define explicitly the objectives and functional requirements of a system. In the context of manufacturing systems, specifications are typically developed around the following areas:

- Processing Machines (CNC machines, robots, manual machines, etc.)
- Material Handling Equipment (AS/RS, AGVs, Conveyers, etc.)
- Operations Control (material flow, order flow, procedures, activities, layout, etc.)
- Organization and Support (direct and indirect labor, skills, computers.)

Within each of these areas individual operational and technical specifications can be developed. Technical specifications are a function of project goals, product features, order quantities, and process technology. Operational specifications are a function of project goals, and internal company practice in order flow, material flow, and various other codes of practice. Without project goals and the identification of particular projects, specifications cannot be developed. However, in the context of creating a road map for manufacturing systems design, it is possible to provide guidelines for the development of specifications and bid specification. The terms specifications and book proposal are used here interchangeably. While specifications make up the central part of the book proposal, other items such as conditions of contract, regulations, and standards are also usually incorporated.

***Bid Specification:*** Once the analysis phase is complete the analysis information generated is used to populate the bid specification. Depending on the specific subsystem being proposed, the book proposal will typically contain sections on the following.

- Company Profile
- Products and Sales
- Corporate and Manufacturing Strategy
- Project Goals and Objectives
- Technical and Operational Specifications
- Standards to be Used
- Health and Safety
- Contract Conditions

- Financial Conditions and Arrangements

- Models (e.g., data flow diagrams, activity models, hardware architectures)

It is important to note that the preparation of a bid specification is only the first step in a process to find the eventual specification for the new system. Bid specifications often change to accommodate supplier constraints and proposed changes. Without the flexibility to change projects, design groups can often find that their project will become very expensive and uncompetitive.

*Development Stages:* Once a project has been understood and analyzed technically by the project group, a number of stages fall quickly into place. These stages include:

1. Development of the operational specifications
2. Preliminary system layout and functional specification
3. Detailed budget costing and capital approval
4. Tendering, selection, and contracts
5. Finalization of operational and functional specifications
6. Contract management and coordination
7. Site management, installation, and testing
8. Commissioning and preliminary training
9. Training, start-up to full operation, and acceptance

The installation and operation of a new system is often a continuous process where incremental change is ongoing. It is important to understand that once a system has been installed, it will be necessary to continue to monitor its performance and make whatever adjustments are necessary, even to replace systems components thought to be properly specified in the first instance.

## 5.7    ROAD MAP

In the preceding sections the IMP procedure has been presented. The application of the IMP procedure is primarily in the domain of systems management and project integration. The primary communication medium used by IMP to facilitate the design group is the IMP model, which is constructed using the IDEFo modeling methodology. The IMP model unambiguously indicates each of the activities in the design process. It also indicates key controls, resources, and flows.

The use of IDEFo in the IMP road map is critical. The IDEFo modeling technique allows us to dissect various stages in the modeling procedure and examine it in more detail. The scope of this dissection is evident throughout the pages of this chapter. However, by using IDEFo it is easy to extend the detail of the IMP road map by simply "exploding" any existing activity of particular interest to a specific project.

The node diagram in Figure 5-15 illustrates the number of diagrams used in the IMP model. Each of these diagrams in turn indicates a number of development activities. The node diagram also shows how the IMP model can be *navigated* by the user who may, for example, only wish to use one activity or perhaps one branch of the IMP model during project implementation. What this node diagram does not show are all the various flows throughout the model for feedback, information and controls. To get this information the user has to read and use the IMP model presented in the preceding pages.

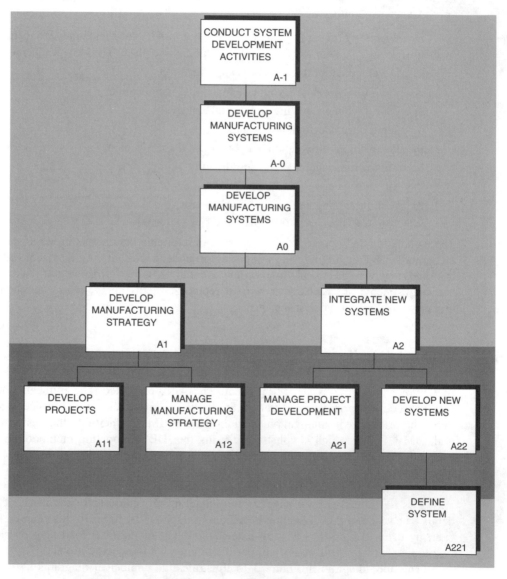

FIGURE 5-15.   Node diagram for the IMP Model

The dark shaded area in Figure 5-15 indicates four diagrams whose activities together form the backbone of the IMP procedure. These activities have been assembled together to form a so called IMP road map for systems design. This road map is illustrated in Figure 5-16. The intention of the road map is to illustrate at a glance the main activities involved in the design procedure and their relationships. Two particular observations can be made about the road map. First, it shows at a glance the many tools, procedures, charts, and checklists that may be used by the design group for augmenting the design process. These *resources* are illustrated using thick black lines for clarity. Second, it shows the control and feedback flows in the design procedure. Control flows are indicated using thin lines and illustrate the types of controls that are applicable at various different stages and more importantly the relationships among these controls. The feedback flows are indicated using dashed lines and illustrate the types of feedback among various different activities. Clearly, the indication of feedback also illustrates that activities are not necessarily executed sequentially but rather in a way which best suits the design group.

## 5.8   CONCLUSIONS

The IMP procedure is the implementable part of the IMP methodology. In this chapter we have seen that the IMP procedure is comprised of an IMP model (created using IDEFo), various, tools, charts, checklists, and the IMP road map. The IMP procedure is said to be an idealized design procedure for the development of projects within manufacturing. From the IMP model it is clear that project development procedures in any environment are potentially complex. IMP allows us to attack at least in part this complexity through the use of modeling techniques for illustrating development stages. It also allows us easier access routes to solving systems design issues by illustrating well-known tools, charts, and checklists from various sources.

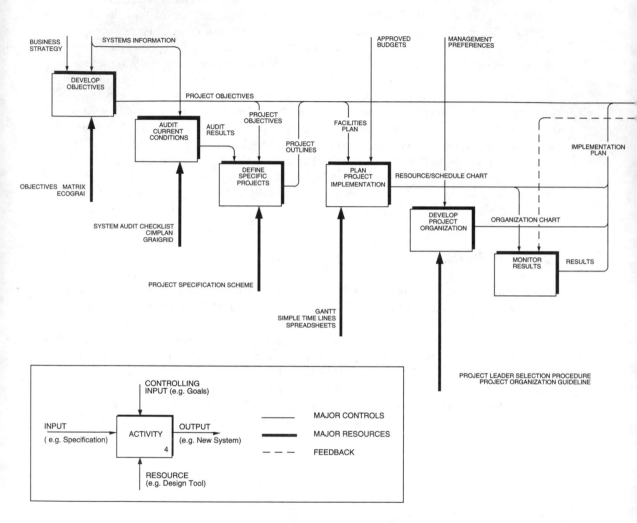

FIGURE 5-16.   IMP Road Map

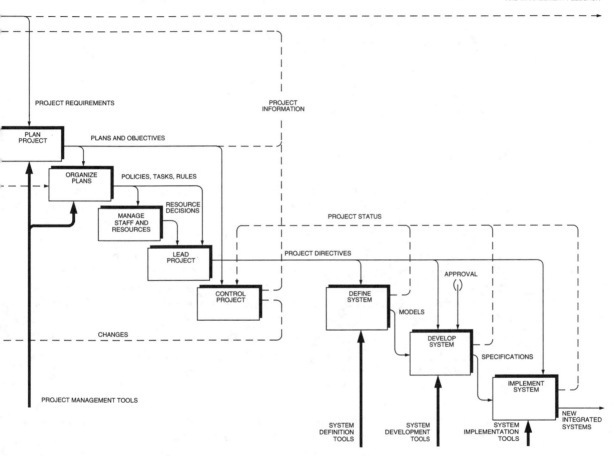

FIGURE 5-16. IMP Road Map (*Continued*)

# Integrated Manufacturing Systems Design

## 6.1  INTRODUCTION

In the previous two chapters we have seen that the IMP methodology contains two parts. The first of these, IMP theory, describes the theoretical foundation upon which the second part, IMP procedure, is built. In this chapter we will look at three case studies which have been implemented with the help of IMP. The three projects presented in each case study use various aspects of the IMP methodology. No one project uses the IMP procedure in its entirety. This reflects on the one hand the flexibility of IMP, and on the other hand the fact that no two manufacturing projects are the same or have the same requirements. However, there are similarities in the way each application uses the IMP methodology.

Each of the three applications have been selected in order to complement each other in presenting each of the five general IMP stages (context development, systems development, projects development, project integration, and project stages). The three applications that have been selected are functionally illustrated in Figure 6-1. Each application has been developed as an independent project in real life. Together they cover three primary areas for development within the manufacturing environment (i.e., production planning and control, processing or machining, and assembly).

Collectively they may also be said to interact with each other as illustrated in Figure 6-1 to form a model of a company which employs all of these functions. Each project is discussed as being independent. But at the same time each is cognizant of the other projects, through the model presented in Figure 6-1. The figure also illustrates which stages of IMP are used in each of the three case studies.

It must be stated at the outset that each of the three applications were implemented during the evolution of IMP, and each have made major contributions towards its development. In this regard, no single application has had the benefit of using IMP from its beginning. Also since IMP is a road map rather than a rigid procedure, the main emphasis for system designers was to use IMP as a guideline and as learning tool.

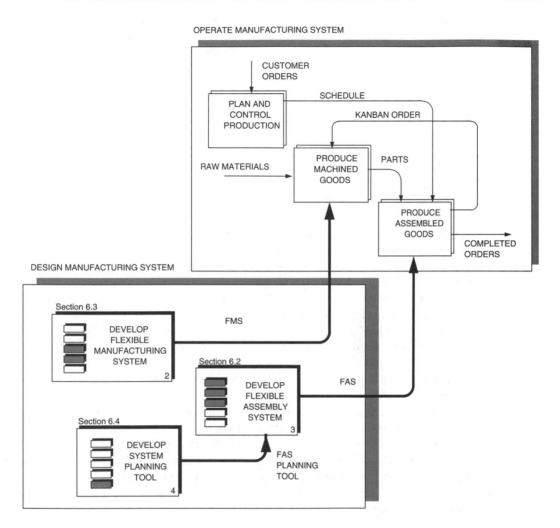

FIGURE 6-1.    Three Applications for Integrated Manufacturing Systems Design

## 6.2    FLEXIBLE ASSEMBLY SYSTEM (FAS)

Manufacturing systems are under increasing pressure to provide a wide variety of products to their customers. In the area of assembly, the increase in the complexity of production makes it more difficult for production systems to remain efficient. To cope with these and other pressures, the production systems need to be designed using the latest technology, methods, and operational concepts. The first case study presented looks at the development of a flexible assembly system that has been developed for a manual flow line assembly process. The system is designed around the production management concept of JIT, and implemented with the support of a software planning system. It has been developed by a large multinational company that has recognized the need for its assembly environment to become more flexible and human centered in order to prepare for the current and future expectations of its customers. The first three stages of IMP (i.e., context development, systems development, and projects development) are discussed.

## 6.2.1 Introduction to FAS Application

In his book entitled "Japanese Manufacturing Techniques," one of Schonberger's seven lessons in simplicity is, *Flexibility opens doors*.[84] This lesson is covered in a chapter that goes on to compare the Japanese approach to the assembly of goods with the approach adopted in the West. In Japan emphasis is placed on addressing the issues of (1) reducing or eliminating set-up time, (2) having flexible labor, (3) foreman level control of line balance, (4) flexible labor assignment, (5) line speed-up/slow-down flexibility, (6) U-shaped or parallel assembly lines, (7) close stations with little or no material handling, and (8) multiple copies of small machines. In contrast many companies in the West place emphasis on (1) line balance (usually carried out by industrial engineers), (2) stability of productions runs, (3) fixed labor assignments, (4) computer support, (5) linear shaped assembly lines, (6) conveyers, and (7) *supermachines*. In short, the Japanese approach of JIT displays a more human centered, flexible system compared to a more structured, computer-based, and inflexible approach in the West. Depending on the assembly system in question, as well as the social and technical environment for the particular company concerned, there can be good and bad features in both approaches.

The project described in this section attempts to marry the best features of the Japanese approach with some of the best features of the western approach. In particular, the Japanese approach to creating human centered flexible systems and mixed-model assembly is used. This is coupled with two computer based tools, one for the design of mixed model assembly systems and one for the planning and control of the assembly system while in operation. The result is the implementation of a system that augments the human skills required to efficiently assembly quality products.

## 6.2.2 Context Development

In general, the assembly function is one of five primary functions carried out on the shop floor. These five functions are (1) Processing, (2) Assembly, (3) Material Handling and Storage, (4) Inspection and Test, and (5) Control. The application described here is concerned with the functions of assembly and control (both material and order control). It also addresses to some extent the function of material handling and storage. Collectively these functions are termed *Control Materials and Assemble Goods* in the IDEFo activity diagram illustrated in Figure 6-2. In this figure three primary activities are shown to be a part of the transactional environment for the activity *Control Materials and Assemble Goods*. These three activities provide the primary upstream and downstream variances for the environment being designed.

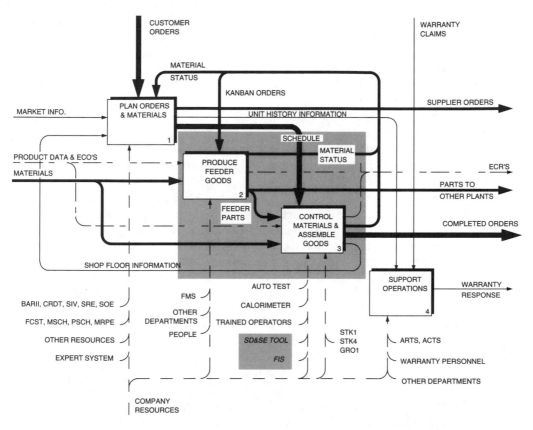

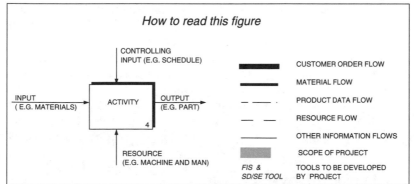

FIGURE 6-2.   Context for FAS Project [1]

---

1.  Some of the terms used in this figure are specific to internal company resources, the description of which is confidential (e.g., STK1).

The first activity, *Plan Orders and Materials* is responsible for providing time-phased schedules for the downstream activities through the use of customer orders, market information, and manufacturing capacity information. The relationship with the assembly activity is through the assembly schedule and material status. The assembly schedule that reflects the assembly capacity and capability lists the type, quantity, and due date of the units to be assembled. In order to assemble these units, materials must be purchased. To carry out this function, information is required on the "stock on hand" from stores. This comes from the *material status* information flow illustrated in Figure 6-2. Although this order flow system is operating relatively well, it was recognized that the Kanban system already in operation in some parts of the factory could be extended to cover the entire plant. The feeder schedule that demanded most of the machined and processed parts could be replaced by a total Kanban ordering system to reduce work in process and improve flexibility.

The second activity *Produce Feeder Parts* represents the activities of sheet metal production, welding, painting, and some subassembly operations. The interaction between this activity and the assembly activity is primarily through the feeder schedule but there is some Kanban material control. One of the key problems in converting over to full Kanban is matching welded frames, which are customer specific, to the parts to be assembled at the beginning of the assembly line. A solution derived from user group meetings is to develop a hybrid ordering system, using Kanban (demand pull) for all common parts and a feeder schedule (demand push) for the special parts. This proposal is enough to begin more detailed analysis and design.

The final activity in the transactional environment for the assembly function is *Support Operations*. This activity represents the support operations of Warranty, Management Information Services, and Manufacturing Systems Design. The primary reason for showing this activity is that a reduction in warranty costs was seen as a primary objective in the redesign of the assembly process. Improvements in quality ratings and workmanship standards will be reflected by a reduction in warranty costs.

In the preceding three activities, which generally model the main sources of variance and key information flows from a project perspective, a number of changes to the manufacturing system were identified from various sources up and down the hierarchy. Although cost reductions resulting from these changes may justify the cost of a project, in most companies there are limited development resources, (i.e., people and funds.) Therefore whatever limited resources are available, must be used in developing projects that adhere to the manufacturing strategy. In other words, projects are often implemented simply because they save money. However, by tying up resources on these projects, other changes that may have a more long-term effect on company competitiveness may be neglected.

## 6.2.3 Systems Development

There are two main reasons for making major changes in any manufacturing facility. The first reason is due to a problem commonly expressed in terms of an increase or decrease in capacity. The second, which is the primary motivation behind this study, is identifying the potential to improve some of the systems performance parameters, particularly productivity and flexibility. Figure 6-3 shows the parameters that have been used in this project as well as a summary of the data obtained during an initial analysis. Also shown in this figure are the aims of the project, which are to improve the various parameters and achieve quantifiable performance values. The vertical axis labeled *assembly system* shows four assembly systems redesigned during this case study. Two of these are multimedia assembly systems, one is a mixed model assembly system, and one is a single model assembly system.

Each of the project parameters were seen as critical to the performance of the assembly system's ability to become more flexible. Adhering to a predefined manufacturing strategy was not practical, since none existed on paper. Therefore, the task of the manufacturing strategist was to glean from users as well as from other managers exactly what they saw as being important, and to try to match this with the overall business strategy which was to improve customer delivery through better flexibility. A number of other issues arising from the business strategy had an impact on the decision of what projects to pursue. The company in question wanted to keep the human centered nature of the assembly area since it judged that human capabilities should be augmented rather than replaced by new technology or automation.

## 6.2.4 Projects Development

The parameters shown in Figure 6-3 are measures of the performance of the existing system. In order to translate these into a set of projects, an analysis phase was conducted to model the existing system, compare it with reference to other systems and the ideal, and then model the target system that would provide the improvements required.

| ASSEMBLY SYSTEM | DEPARTMENT | UNITS PRODUCED | STD HRS/UNIT | AVG HRS/UNIT | INVENTORY | SUPERVISORS | EXPEDITERS | AVG QUALITY RATING | AVG SET-UP LOSS | SPACE UTILIZATION FACTOR % | MATERIAL HANDLING FACTOR | ON-TIME DELIVERY | MANAGEMENT INFORMATION | BALANCE DELAY |
|---|---|---|---|---|---|---|---|---|---|---|---|---|---|---|
| MULTI #1 | A | 907 | 15 | 30 | 300 | 1 | 1 | 787 | 930 | 31 | 9 | 57% | | 24% |
| | B | 114 | 14 | 50 | 150 | | | - | | | 9 | 63% | | 32% |
| MULTI #2 | C | 173 | 10 | 19 | 65 | 1 | 1 | - | 2592 | 63 | 9 | 98% | | 34% |
| | D | 596 | 21 | 37 | 90 | | | 764 | | | 9 | 90% | | 26% |
| SINGLE | E | 4613 | 26 | 39 | 35 | 2 | 3 | 742 | 192 | 64 | 12 | 78% | | 24% |
| | F | 188 | 33 | 60 | 26 | | | - | | | 12 | 90% | | 35% |
| MIXED | G | 1593 | 22 | 34 | 47 | 1 | 1 | 822 | 0 | 59 | 9 | 98% | | 35% |
| | H | 663 | 22 | 40 | 38 | | | 731 | | | 9 | 97% | | 23% |
| AIMS | | | | ↓10% | ↓25% | — | ↓6 | ↑900 | ↓20% | ↑60% | ↓ | ↑100% | ↑ | ↓15% |

FIGURE 6-3.    Project Performance Parameters

Using a combination of brainstorming, factory visits, and research, three projects were ultimately identified which together could improve each of the parameters. The mapping of the parameters onto each of the projects is illustrated in Figure 6-4. The three projects were (1) JIT Material Control, (2) Cell Design and Evaluation Tool, and (3) Factory Information System, brief models of which are discussed in the section on project operation.

Although they are shown simplified in Figure 6-4, the stages carried out at the systems development level in identifying the three projects and having them justified was quite detailed, as indeed were the models created to facilitate analysis. The various stages can be summarized as follows: (1) development of project objectives, (2) audit of the existing system, which included evaluation and forecasting of requirements, (3) definition of specific projects, including defining alternatives and carrying out a financial analysis and selecting alternatives to be pursued, (4) planning project implementation, giving general time scales, resource requirements, and finances, (5) developing the initial project team structure, and (6) monitoring the results of each project implementation.

## 6.2.5 FAS Conclusions

The first three levels of the IMP methodology have been used in the initial development of a flexible assembly system. The theoretical part of IMP allowed the project team to view their environment in terms of the seven principles outlined in Chapter 4. It also allowed them to review some ideas documented in various texts by well-known manufacturing specialists. The outcome was a prepared work group capable of understanding the concept and limitations of their environment holistically.   It is difficult to

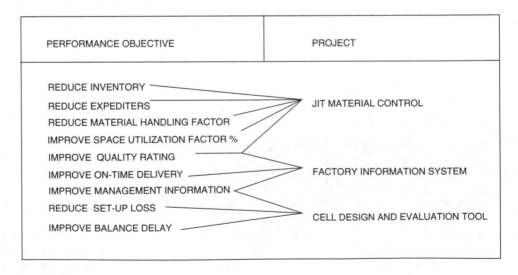

FIGURE 6-4.   Three Projects

quantify the actual benefits that the theoretical part of IMP had on the project, but what is clear is that the group as a whole was realistic about setting goals, examining the openness of their system, and ensuring that social and technical considerations were discussed equally.

The procedural part of IMP outlined a series of stages both generic and specific for the project group to undertake. Following goal definition and project team organization, the context for the project was clearly mapped using some structured approaches. In addition, user requirements were matched with management goals. The result was a set of definable and implementable projects which achieved the match between goals and user requirements. It was clear during each of these preliminary stages precisely what action was required next in terms of mapping goals with actual projects, and precisely what actions were to be scheduled and implemented. The IMP procedure facilitated this by providing the road map on what activities were to follow, how these activities could be resourced, and the types of information and controls that interlinked each activity. More information is available on this application through various publications associated with this research.[85, 86, 87]

## 6.3 FLEXIBLE MANUFACTURING SYSTEM (FMS)

A Flexible Manufacturing System is a complex association of computer-controlled equipment, automated material handling systems, computer-control software, and people. The development of such an association can be elaborate, requiring experts from a number of functional disciplines who, when brought together, must operate as a fully integrated design group. This group in order to cope with the many operational and organizational difficulties that often arise, must be supported by tools and methodologies that can reduce the effects of these difficulties and make the whole design process more efficient. The operation of the system must be designed to reflect state-of-the-art material flow concepts and process technologies. The application described in this section deals with the development and operation of a sheet-metal flexible manufacturing system, which has been implemented using structured tools and methodologies and uses the demand-driven mode of material flow scheduling. The effects have been to reduce design time, improve implementation times, and maximize efficient operation. The case study deals mainly with the third and fourth stages of IMP (i.e., project development and project integration).

### 6.3.1 Introduction to FMS Application

A Flexible Manufacturing System (FMS) may be considered to be a miniature factory within the broader context of a manufacturing enterprise. In Figure 6-5 the activities of an FMS are shown in relation to the other activities in the enterprise. Both the FMS and operation planning and control boxes in this figure correspond to the general activity of production operations, which encompass all those activities necessary for converting raw materials into finished products. The FMS and operations planning and

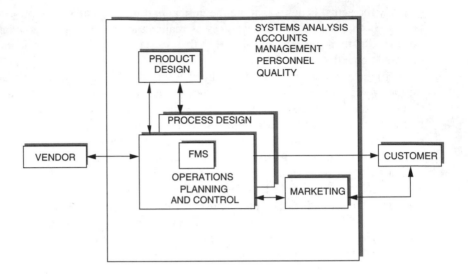

FIGURE 6-5.    Context for FMS

control boxes are shown layered to each other to reflect the considerable amount of interaction between the two activities.

The operation of an FMS cannot take place in isolation. As shown in Figure 6-5 a number of functions interact with the FMS directly on a day-to-day basis. Two of the most important functions that interact with the FMS are Design and Operations Planning and Control. The interactions between the FMS and Design will be described under the terms CAD/CAM and Computer Aided Process Planning (CAPP).

*CAD/CAM:* The term CAD/CAM can have a number of meanings. In this section the term CAD/CAM refers to the interaction between the CAD systems of Design, NC part programming systems, and the FMS controller. Figure 6-6 shows the relationship between these systems for a sheet metal FMS. In this figure the CAD system uses a minicomputer and applications software from a leading CAD supplier. The NC part programming system also uses a minicomputer and applications software, but from different suppliers. The reason for this is that both areas evolved as islands of automation utilizing the best systems available at the time for the intended functions. To communicate graphic drawings between these two graphic systems, the file transfer mechanism IGES was installed. Transfer of data from the NC part programming system to the FMS was performed using a standard file transfer program, since only the NC part program needed to be exchanged. An important point in this configuration is that the Design function (through its CAD system) and the manufacturing function (through its NC Part programming system), have clashed regarding integration of the graphic systems. This clash has been somewhat alleviated through the availability of the IGES software for both systems. Issues of training, different operating systems, maintenance contracts, shared storage, etc. are not so easily addressed.

*CAPP:* The activity of generating process plans for the sheet metal components is assisted by the use of a generative process planning system. In this system the user, or process planner, enters various parameters in relation to the new part (e.g., material thickness, process feature) and is prompted by the process planning system into entering new data or accepting the data recommended. The result is a process plan for the part in terms of the processes that the part must pass through and in what order. The actual machines used to carry out the process are selected by the FMS controller based on availability so as to optimize the entire system.

*FMS Controller:* The FMS controller can be divided into two important areas: Hardware Architecture and Software Architecture.

*Hardware Architecture* A hardware architecture for the sheet metal FMS is illustrated in Figure 6-6, which shows a computer network that offers room for system expansion and speed of data transfer. The topology is a "bus" structure, which enables machines and other equipment to be "plugged" into the system at the nearest convenient point. There is extensive use of multiplexers which communicate with the equipment using standard low-level communication protocols. As can be seen from Figure 6-6, the CAD system and the FMS controller can both communicate with each other using the network. The NC part programming system is not linked to the network and only communicates using the standard low-level communications protocols. It is anticipated that the CAD system will eventually replace the activities of the NC part programming system, as soon as the machine interfaces are fully integrated and the same functionality is achieved.

*Software Architecture* The software architecture utilized by the FMS controller is illustrated in Figure 6-7, which shows the various software modules hidden behind functional boxes to better illustrate the operation of the FMS controller. The planning module is responsible for taking in orders from any of three sources. These orders are grouped according to a set time frame or time window. Once a group of orders are received, they are reviewed if necessary and passed on to the preproduction scheduling module, which checks for material availability and rejects orders requiring out-of-stock material. Once an order has been accepted, material is reserved for it. The preproduction scheduler creates the optimum schedule based on various scheduling factors and resources available. This schedule is translated into a set of near-term production schedules or nests. This nesting program, because of the nature of the production management control system (i.e., JIT and Kanban), is dynamic and only creates nests based on actual orders released by the preproduction scheduling module. The cell planner reserves at all times discretion on whether to release nests for production. When the nests are accepted, they are passed to the production module which sends the machine instructions to each individual machine.

*Production Management:* Two important production management areas require specification with respect to the operation of a sheet metal FMS. These are JIT and Kanban, and Shop Floor Control.

*JIT and Kanban* In a manufacturing environment evolving toward full-blown JIT and Kanban order entry, the choice of order control mechanism for the FMS is relatively

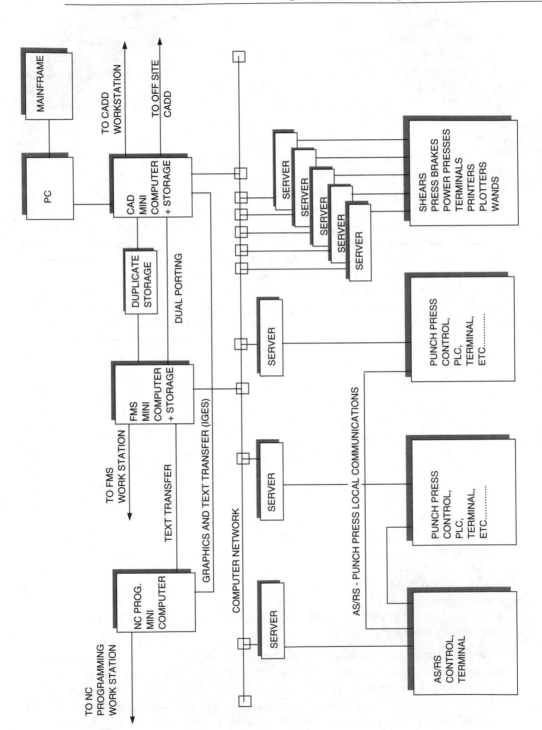

FIGURE 6-6.    Hardware Architecture

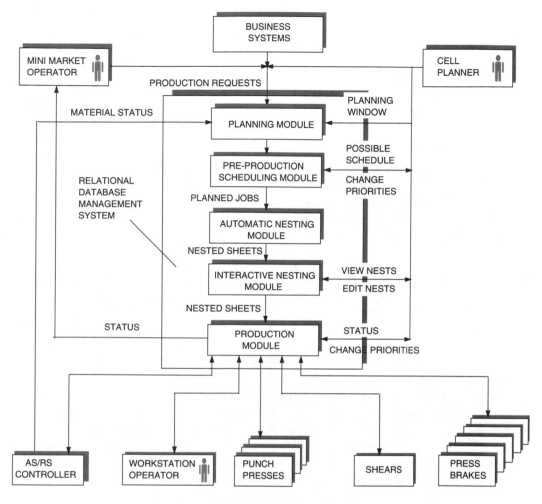

FIGURE 6-7.   Software Functions

easy. The Kanban system of order entry is particularly well suited to repetitive production systems and contributes to reducing inventories and providing increased flexibility to part order and design changes. The operation of the Kanban system in its simplest form is illustrated in Figure 6-8.

In the Kanban system orders are placed into the system by downstream activities to the FMS (e.g., Assembly) through the use of production Kanbans and bar code technology. These Kanban tickets arrive at the FMS minimarket in an empty box. The Kanban is wanded and the order placed automatically into the FMS order list. The quality of parts to be produced and the cycle time are stored in the database and occasionally adjusted by the industrial engineering function to either reduce further the cycle time or improve any shortage problems. The Kanban card is stored by the "kitter." When the

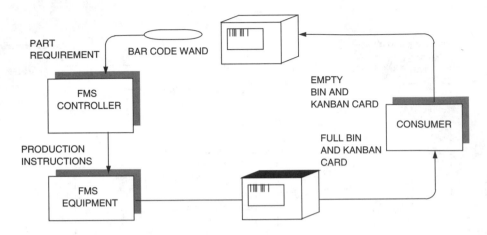

FIGURE 6-8.    Kanban Order Entry

part is processed on the first manually operated work station, a new production Kanban is printed to reflect any new Kanban details (e.g., revision changes, new movement details, etc.) and is placed with the parts. The production Kanban doubles as a conveyance Kanban or move ticket and the parts are moved through the system. When the parts arrive at the kitting area the new Kanban is wanded so that the FMS controller can take credit for their production, the old Kanban destroyed and the parts moved to their next destination (e.g., Assembly).

*Shop Floor Control*  Various parameters are monitored in the system to allow the system to be effectively controlled. The use of bar coding at each of the processing stations allows order progress to be monitored. Each of the machines in the system is also monitored for various machine specific parameters (e.g., utilization, set-up time, idle time, maintenance times, etc.). These monitoring devices are manually activated by the operator at the machine by pushing certain buttons for each machine status. Getting this information in any other way would be extremely difficult. Each operator also has a video terminal for receiving and giving information to the system. The terminal is also used to monitor problems experienced by the users at the various work stations.

## 6.3.2 Project Integration

In the past, the duration of project planning has been relatively short with designers preferring to adopt an aggressive and prompt implementation of the project. The organization of the project has typically involved only one function, either manufacturing systems design or MIS with cursory exchange of ideas and information by various project liaison officers. In the case of manufacturing systems, the manufacturing department has taken exclusive control over the planning and execution of the project. In today's environment complex projects such as the development of an FMS involve more sophisticated equipment, have a bigger impact on the manufacturing environment, and cross many functional boundaries not only at the operational level (e.g., production, materials, and shipping) but also at the planning level (e.g., management information systems and manufacturing planning). To cope, these projects require more structured planning methodologies coupled with a more tolerant learning approach in their planning phase. They also require management-led project teams involving all functions affected by the project.

*Project Organization:* Various project organizations have been described in previous chapters. In the development of an FMS, the project team will typically be organized as a multifunctional group. In this structure the project team is led by a senior business manager, directly affected by the implementation of the project. This manager is responsible for planning, organizing, staffing, leading, and controlling each aspect of the project from the social aspects of project team productivity to the technical aspects of project development tools and implementation of the project onto the shop floor. The type of internal organization and interaction within the group will hinge primarily on the managerial talents of the project manager. In this project the "heavy-weight" project team was chosen.

*Structured Methods:* Structured methodologies allow designers to share a common perspective on a proposed systems functionality. In the past, structured methodologies have been primarily used in the development of software systems and in areas of business requiring management of large quantities of information. However, as the manufacturing system is being increasingly looked on as a complex information system, these structured methodologies are increasingly used in the development of large and small manufacturing projects. The advantages of using structured methodologies are clear. As mentioned earlier they promote a common perspective for designers who may come from different functions. They also allow unambiguous definition of various specifications which by and large go to make up software systems developed by outside subcontractors. A large number of structured methodologies now exist for the development of systems such as an FMS. In the development of the sheet-metal FMS the following analysis tools and methods were appropriate and contributed to the effective implementation of the system.

1. Group Technology/Production Flow Analysis, GT.

2. Functional Modeling Methodology, IDEFo.

3. Data Flow Modeling Methodology, SSA.

4. Simulation/Animation Modeling, SIMAN/CINEMA.

The production-flow-analysis technique from Group Technology (GT) is a general analysis tool for the detailing of machine layout and the creation of GT cells. Functional modeling may be considered to be one of the first system modeling procedures carried out by the project team and used for definition of the functional specifications. Data flow modeling is a lower stage of modeling by which the functional models are converted into data flow models for subsequent translation into program codes where appropriate. Finally, simulation and animation modeling is an effective tool for testing the dynamic behavior of the system prior to final specification.

*Performance Parameters:* Many other factors affect the development of an FMS. Factors are seen as critical for the control of the development of a project of this nature is the performance parameters by which the project's success can be measured. The performance parameters will depend on each company's particular manufacturing strategy and the critical success factors important to operations. The way in which the parameters are stated and formulated must reflect a common view and perspective involved with the project. Equally important, these parameters must be capable of unambiguous measurement and control. A list of the important performance parameters used in the development of the sheet metal FMS are given.

| Hours/ Unit | Scrap Rate | Work in Progress |
|---|---|---|
| Capacity | Cycle Time | Rejects Quantities |
| Inventory | Man Utilization | Set-up times |
| Shortages | Machine Utilization | Maintenance Costs |

Clearly, some of these parameters are dependent on others and indeed all can be expressed in terms of the rate of return for the investment. However, these parameters, although requiring additional definition, can be measured before and after project implementation and used as the primary control on project performance.

## 6.3.3 Review of Selected Parameters

A considerable amount of effort often goes into the justification of high-technology projects. Much of that effort is concentrated on convincing top management that the proposed project is viable. With projects of this type, management must take strategic decisions and this involves an element of risk. It is thus vitally important that the same level of effort go into carrying out a post-evaluation exercise. This is necessary not only to evaluate whether the original objectives were met, but also to allow the project team to evaluate their original decisions. In the subsections that follow, a number of measurement parameters are presented for the sheet-metal flexible manufacturing system.

*Capacity:* One of the major impacts of the FMS implementation was to increase the sheet-metal manufacturing capacity in line with product demands. The old manufacturing methods could not cope with this increased demand, which resulted in a high level of subcontracting. The tracking and monitoring of subcontract levels is important to control costs and monitor the effectiveness of reducing it. Figure 6-9 illustrates this subcontract level. The graphing of the levels of subcontracting over a period of three to four years showed many high peaks. An analysis of these peaks revealed that they coincided with the time during which new products were being put into production. It showed that the old system's were unable to cope with the sudden demands of new product integration. Thus, subcontracting was used as a relief valve to get over this short-term hurdle, which proved to be a poor way to control the manufacturing system. The new system had little difficulty coping with these peak demands as once off, prototype parts could be manufactured in a very short time before being released into production.

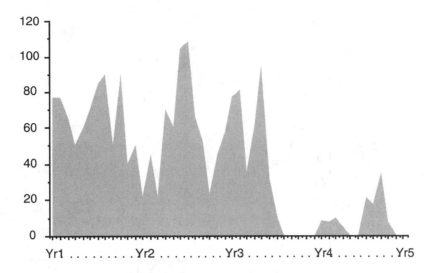

FIGURE 6-9.    Sheet Metal Subcontract

As new equipment was installed, the ability to manufacture in-house parts increased accordingly. Figure 6-10 is a plot of demand against in-house manufacturing versus performance over the duration of the project. The difference between the two curves is the level of subcontract and this is at its highest level in the early years as shown in Figure 6-9. The system in its present configuration is coping with the demand based on a 15-shift operation per week.

*Productivity:* One of the major difficulties with the old system was that the level of labor was increasing at the same level as product demand. This meant that costs were not being controlled and the overall sheet metal product cost was increasing rather than decreasing, as volume increased. A high level of operators' time spent was also spent on material handling. The FMS showed some dramatic improvements in this situation, as illustrated in Figure 6-11. Using the technology introduced by FMS, the number of units produced per person increased threefold.

*Manufacturing Cycle Time:* In the old manufacturing system, an average of two and one-half month's supply of parts were on hand at any one time. The old shop order system would instruct up to three month's supply of many parts to be manufactured at any given time. This required high raw-material inventory levels and high work in process levels. Figure 6-12 shows the improvements that came about as a result of introducing the FMS. Cycle time is currently down to five days in line with the original set target. This shorter cycle time has a major impact on the performance of the system when it comes to responding to increased customer demand.

*Cost Time Profile:* When raw material is brought into the plant, work is carried out on this material such that value is added to it. In the case of sheet-metal, this value added includes holes being punched and parts being formed, thus transforming it into work in

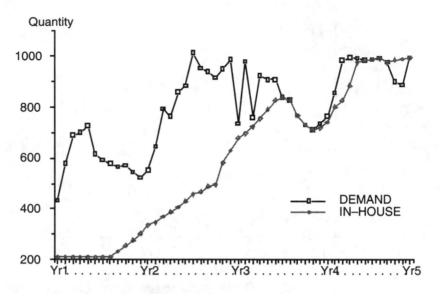

FIGURE 6-10.    Comparison of In-house Production and Demand

process. An analysis was used to calculate the value added to this raw material during the manufacturing cycle time. This was done by taking the total amount of process time and multiplying it by a cost-per-hour factor for the individual machines. A plot made of the money value issues per week of inventory against time gives what is termed a Cost-Time Profile. Figure 6-13 illustrates the values involved for the sheet-metal fabrication department. There is a two-week cycle time involved for raw material as the majority of it comes from Europe. The in-house manufacturing cycle time is then added, giving a three-week cycle time from raw-material order to consumption in assembly. This compares with a cycle time of over seven weeks for the old system. The new system had a squeezing effect on this overall cycle time. The issues per week increased due to the increased output from the system. The total inventory level thus decreased slightly between the start and the end of the project. This must be put in perspective as the overall throughput increased by 300% over the same period. The dotted line thus shows what the level would be if no improvements were made.

## 6.3.4 FMS Conclusions

Clearly this project was highly complex in nature. The overall cost of the project reflected this. The IMP methodology rather than imposing a strict sequence on the implementation of the project acted as a guideline and educational tool for the project group. As in the first case study, the theoretical part of IMP allowed the project group to appreciate the key aspects of systems development in a general sense. The project team could understand that systems are goal seeking, have social and technical subsystems, and are learning organizations. This created a reference marker for the group, and dur-

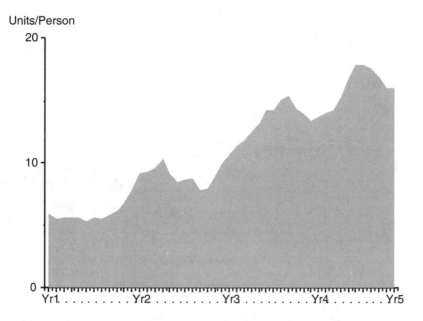

FIGURE 6-11.    Manpower Productivity

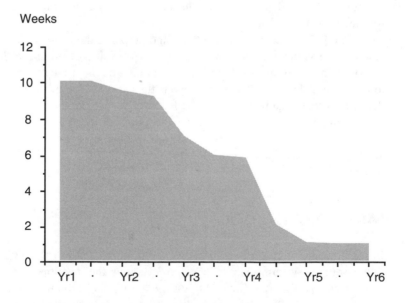

FIGURE 6-12.    Manufacturing Cycle Time

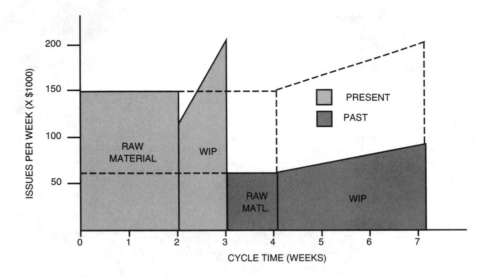

FIGURE 6-13.    Cost Time Profile

ing the implementation of the project, the IMP principles were continuously returned to, reiterated, and tested in terms of the then current direction of the project. If the group felt that any one of the key principles were not being addressed by current activities, then the emphasis would change to activities that did. These activities tended to be more strategic than routine and hence essentially more important for the future performance of the project.

The IMP procedure on the other hand prepared the project group for the enormous amount of routine activities that were pursued. Some of the initial Project Integration activities involved the development of detailed models of the operational activities. Due to its size a major issue in the project was the organization of the project group. Initially the project was organized around the lightweight project manager concept (see Section 3.4.1) but eventually after following the guidelines illustrated in the IMP model, it moved toward a more heavyweight project manager organization. Many other issues arose, and these have been documented in various publications related to this research.[88, 89, 90, 91]

## 6.4    FAS PLANNING TOOL

In the first case study three projects were outlined as part of the development of a flexible assembly system. One of these projects was the development of a so called Cell Design and Evaluation Tool. This tool was to be a computer-based decision support system for the assignment of operations to cells incorporating both human and robotic resources. In this case study the development of such a tool in more detail will be described. The tool, entitled OPERAT, has been developed a part of as larger project entitled CIM-PLATO (cf. Section 2.6). This case study deals mainly with the final stage of IMP (i.e., project stages).

## 6.4.1 Introduction to FAS Planning Tool Application

The overall objective of the CIM-PLATO project is the development of an industrial toolbox prototype of computer-based procedures and tools, which support the design, planning, and installation of FMS and FAS systems in a CIM environment. The proposed hierarchy of tools has been presented earlier in Section 2.6.

## 6.4.2 Project Stages

The project was developed around three steps as illustrated in Figure 6-14. Each of these three stages was supported with structured analysis and design tools. Each of the tools selected adopted a hierarchical and holistic approach to software development. In the first step, the structured analysis tool of IDEFo was selected to generally model the

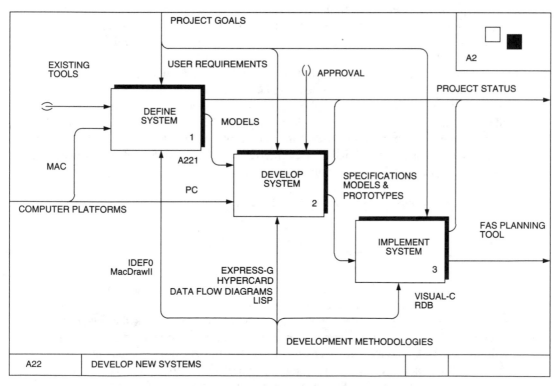

FIGURE 6-14.   Development Activities

activities and process of the software tool. In this process IDEFo models were created, which reflected on the one hand user requirements, while on the other the constraints of project finances, project technology, and innovativeness over other existing tools on the market. Figure 6-15 illustrates one activity developed using the IDEFo methodology and used as a requirement definition for the next two steps of software development.

The second step of project development used a three-pronged development approach for the development of software functionality, data structure, and user interface. The development tools adopted for software functionality were LISP and Data Flow Diagrams. Again, these tools adopted a hierarchical and holistic perspective on tool development. LISP allowed the rapid prototyping approach to be used to prove software functionality prior to eventual coding using Visual-C. Data Structure design was facilitated using the EXPRESS methodology, again a hierarchical tool. Finally, user interface development was facilitated through use of the current de facto standard on user interfaces, OSF-Motif and a versatile graphics processor. While the standard illustrates many models for icon, menu, and window design, the graphics package allowed the design group to prototype the interface, initially using dummy text. Later, a dynamic prototype was created using Hypercard, which showed users how the eventual software would look and feel. Figures 2-3, 2-16, and 6-16, illustrate three models created using each of these techniques. These models were used by designers in the final stage of tool development.

The third and final step of tool development was Implementation in which the tools were more applications oriented. In place of LISP and DFD, Visual-C was implemented for enhanced flexibility and performance of the software tool. EXPRESS led to file structures implemented using a combination of Visual-C and SQL for a database management system. Finally, OSF-Motif and Hypercard models were used in the creation of the final coded Visual-C user interface. All three series of codes were then linked to create the final functioning tool.

## 6.4.3 FAS Planning Tool Conclusions

Unlike the other two case studies, this project involved the use of only a relatively small element of the IMP methodology. The theoretical part of IMP facilitated the project not only by providing a vision for the project team in terms of goals and user requirements but also in encouraging the group to adopt a broad range of analysis tools for holistic and hierarchical systems analysis.

With respect to the use of the IMP procedure, it is clear that the lower levels of procedure were applicable in this project as goals were already well defined and project groups selected. Three stages were identified for project development in accordance with the IMP road map. Through these stages the various inputs, outputs, and resources could be readily identified. More information is available on this application in the various publications associated with this research.[85, 86, 87, 92]

The three case studies that have been presented have used the IMP methodology to varying degrees during their development. The benefits gained by each project are

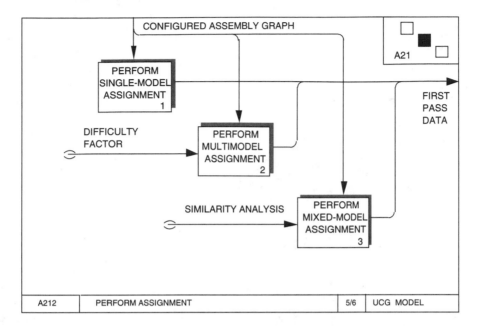

FIGURE 6-15.    Sample IDEFo Model for OPERAT

clear. IMP provided an unambiguous road map for the development activity, which facilitates the design process through better management, integration, and implementation. The actual benefits however, were not maximized as the IMP methodology itself was being developed in parallel to the development of each project. Indeed, the interaction between project development and IMP development became a two way learning process. On the one hand, many of the ideas that are embedded in IMP were tested, validated, and altered through feedback from each of the projects during use. On the other hand, a number of sound ideas were channeled into each of the projects, which ultimately led to increasing the efficiency of their implementation and their understanding by the project group.

## 6.5    THE FUTURE MANUFACTURING ENVIRONMENT

There is little doubt that the current manufacturing environment is changing in order to meet the increasing demands of customers globally. The effect of these changes is challenging systems designers to continuously develop manufacturing systems for becoming the competitive weapons of industry. In light of the rapid changes that have occurred in eastern Europe, keeping abreast of changes is often difficult for company strategists. In Chapter 1 the survey conducted by the Society of Manufacturing Engineers outlined many of the changes facing manufacturing engineers as they move into the 21st century. It can be argued that their conclusions reflect the need for a systems

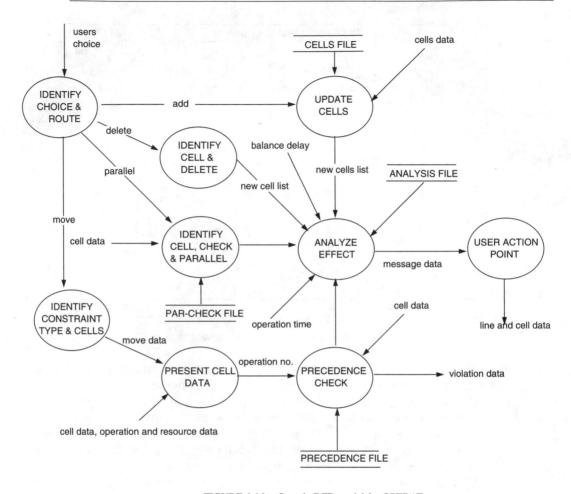

FIGURE 6-16.   Sample DFD model for OPERAT

approach by engineers and systems designers toward their work. This systems approach is echoed in a widely read report from Browne, Sackett, and Wortmann who describe the need for a *total manufacturing business "systems" development.*[93] Their report, commissioned by the European Commission, identifies potentially new areas of research in manufacturing. They outline many attributes of the manufacturing system of the future and in particular the main areas for development in the context of manufacturing research. In this section we shall look at the results of their findings and how they may impact on the environment of manufacturing systems designers in the next century.

Three main factors will cause existing manufacturing environment adapting and changing to meet the demands of the next century. These factors are globalization, environmentally benign production, and changing organizational structures. Each factor will ultimately have its own effect on how the manufacturing system will adapt. Together they will change the essence of manufacturing as we know it today from one of functional excellence and efficiency to one of inter-enterprise integration. Globalization reflects the ever-increasing trend for manufacturing companies to trade long distances away from their traditional markets. In the past relatively few, albeit very large, manufacturing enterprises could afford to operate in distant markets. In the future even the very smallest manufacturing enterprises may need to trade globally in order to remain competitive. They will increasingly source their materials from all corners of the globe and in return find new markets for their products. The net effect of globalization is that in order to product cycle time more pressure will be placed on reducing manufacturing lead times and the value that can be added during manufacturing.

The rapid changes in globalization and the increase in manufacturing activity in general is also reflected in damage to the natural environment. Many governments from the more progressive industrial nations of the world will be asking industry to manufacture products that are friendly to the environment. The changes necessary to make products more environmentally friendly will include sourcing newer materials, manufacturing products that are biodegradable and suitable for recycling. In this scenario manufacturers will not only be responsible for designing and producing products efficiently, but may also be responsible for their eventual recycling and waste handling. Even now some countries have required companies to buy back old equipment, such as refrigerant-cooled computers and refrigerators, and to properly recycle the harmful materials.

The net effect of these pressures will be that the traditional manufacturing facility as described in this book will increasingly be seen as merely a single cog in a longer value chain. Indeed the value added to product in manufacturing, through physical transformation will diminish and perhaps even reach zero. The longer value chain, which includes the suppliers, manufacturing, distribution, and sales, will be responsible as a whole for increasing the value of product and issues such a improved service, shorter lead times (necessitating the integration of suppliers, etc.), and better marketing will take precedence over the more manufacturing related issues. In summary the manufacturing facility will be a small part of a larger chain of activities, and the key design issues will tend to be related to the integration of enterprises.

Inter-Enterprise integration (depicted in Fig. 6-17) essentially entails compressing the time required to bring a new concept to the customer. The production management paradigm is one where real-time orders from customers drive each activity in manufacturing from design through materials planning to shop floor control. This contrasts with the traditional paradigm where products were conceived and manufactured for stock, and the current paradigm where products are conceived by the company and then manufactured to customer orders. In the future scenario, customers will literally decide their specifications in advance and receive the finished product within a reasonable time frame. Key technologies such as electronic data interchange, communications, and information processing will dominate the manufacturing strategy and subsequently the main systems design activities.

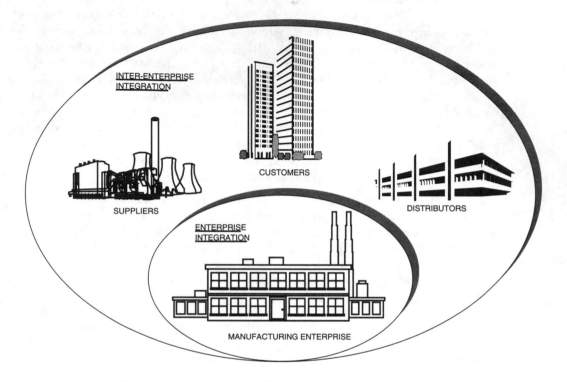

FIGURE 6-17.   Inter-Enterprise Integration

## 6.6    CONCLUSIONS

IMP is a new methodology for the integration of manufacturing systems into their technical and social environment. In this book, the IMP methodology has been described as consisting of two essential parts—IMP theory and IMP procedure. The theoretical part of IMP is a synthesis of well-known theories in systems design, including General Systems Theory, Sociotechnical Design, Learning Organizations, and Structured Analysis. The procedural part of IMP is built around the IMP road map, which guides systems design groups through various stages in systems design, highlighting key tools and resources as well as important information flows.

A major feature of the IMP theory is that it encourages designers to create a view of manufacturing that is consistent with many other well-known theories. In particular, the IMP theory:

- Promotes the use of general and open systems theory.

- Brings the concepts of general systems theory closer to manufacturing systems.

- Promotes the idea of systems designers using ideological benchmarks (e.g., GST).

- Synthesizes the concepts of open systems theory and sociotechnical design.

- Synthesizes the concepts of open systems and learning organizations.

- Creates an *easy-to-remember* reference theory for systems design.

- Supports state-of-the-art issues through the Ideas for Change.

- Promotes its own learning through changes of these Ideas.

The IMP procedure has some similarities with existing systems design tools, since the latter make prime use of the concept that project goals are derived from a business strategy and must be decomposed down through the various stages of project development. IMP however has a number of important additional features. First, it is derived directly from the IMP Theory. In this respect, IMP is designed to promote many theoretical concepts. A second difference is that IMP recognizes the use and application of technical and social planning tools during the implementation stages. The IMP procedure identifies popular tools and techniques and the stages in systems design where they can be applied. Additionally, because the development process like the manufacturing process, is in itself complex, the IMP procedure uses IDEFo to facilitate structured and easy-to-handle information for the designer. In general, the IMP procedure:

- Is developed around a well-defined systems design theory (IMP Theory).

- Uses the IMP theory throughout the IMP model.

- Uses IDEFo for unambiguously describing project implementation activities.

- Decomposes goals and user requirements throughout the design process.

- Promotes the use of checklists and charts for better project management.

- Presents a holistic approach ensuring that no major issues fall between the *cracks*.

- Is easy to understand yet transparently contains major systems concepts.

- Guides designers without prejudice to their own thinking.

- Allows complex design procedures and issues to be handled in small *chunks*.

- Promotes *learning* among the systems design group.

The present complex nature of systems, as well as the changing social environment presents unprecedented challenges for the manufacturing systems design group. The need for new approaches and tools that can facilitate the design group are essential for the creation of truly integrated manufacturing systems. New tools must be capable of addressing a number of issues in integration, including social and technical issues. Commentators agree that no tool currently exists that can address both of these areas holistically. While many tools address individual and smaller issues within each area, these tools are not linked. The new IMP methodology addresses both social and technical integration issues. IMP provides the systems designer with an approach to systems design that is truly holistic and capable of reaching the outer limits of productivity through integration.

# Glossary

| | |
|---|---|
| **AMRF** | Automated Manufacturing Research Facility |
| **CAD** | Computer Aided Design |
| **CALS** | Computer Aided Logistics Support |
| **CAM-I** | Computer Aided Manufacturing—International |
| **CASE** | Computer Aided Software Engineering |
| **CIAM** | Conceptual Inferential Abstract Modeling |
| **CIM** | Computer Integrated Manufacturing |
| **CIM-OSA** | CIM—Open System Architecture |
| **CIM-PLATO** | CIM—Systems Planning Toolbox |
| **CNC** | Computer Numerical Control |
| **CORE** | Controlled Requirements Expression—activity modeling method |
| **DEC** | Digital Equipment Corporation |
| **EDIF** | Electronic Data Interchange Format |
| **ESPRIT** | European Strategic Program of Research in Information Technology |
| **EP ####** | Esprit Project #### |
| **FA** | Functional Analysis |
| **FAS** | Flexible Assembly System |
| **FMS** | Flexible Manufacturing System |
| **GIM** | GRAI Integrated Method |
| **GKS** | Graphics Kernal System |
| **GRAI** | GRAI—decision modeling methodology |
| **GST** | General Systems Theory |

| | |
|---|---|
| **GT** | Group Technology |
| **HP** | Hewlett Packard |
| **IBM** | International Business Machines |
| **ICAM** | Integrated Computer Aided Manufacturing |
| **IDEF** | ICAM Definition Methodology |
| **IEM** | Information Engineering Method |
| **IGES** | Initial Graphic Exchange Specification |
| **ISAC** | Information Systems Analysis and Control |
| **JSD** | Jackson System Development |
| **IDEFo** | ICAM Definition Language Zero |
| **ISO** | International Standards Organization |
| **ISP** | Information Strategy Planning |
| **ISSM** | Information Systems Specification Methodology |
| **JIT** | Just in Time |
| **JMA** | James Martin Associates |
| **JSD** | Jackson System Development |
| **MAP** | Manufacturing Automation Protocol |
| **MRP** | Materials Requirements Planning |
| **MIS** | Management Information Systems |
| **MMS** | Manufacturing Message Specification |
| **NBS** | National Bureau of Standards |
| **NIAM** | Nijssen Information Analysis Method |
| **OSI** | Open Systems Interconnect |
| **PDES** | Product Data Exchange Specification |
| **PLC** | Programmable Logic Controller |
| **SADT** | Structured Analysis and Design Technique |
| **SASD** | Structured Analysis and System Design |
| **SD** | Sociotechnical Design |
| **SDM** | System Development Methodology |
| **SMT** | System Modeling Tool |
| **SPT** | System Planning Tool |
| **SQL** | Structured Query Language |
| **SSA** | Structured Systems Analysis |

| | |
|---|---|
| **SSADM** | Structured Systems Analysis and Design Methodology |
| **SSM** | Structured Systems Methodology |
| **TOP** | Technical Office Protocol |
| **USE** | User Software Engineering |

# References

1. Skinner, W. 1985. *Manufacturing—The Formidable Competitive Weapon*. New York: Wiley.

2. Koska, D. E., and J. D. Romano. 1988. *Profile 21 Issues and Implications*. Michigan: Society of Manufacturing Engineers.

3. Jaikumar, R. 1986. Postindustrial Manufacturing. *Harvard Business Review*, 37:79-86.

4. Hayes, R., S. Wheelwright, and K. Clark. 1988. *Dynamic Manufacturing—Creating the Learning Organization*. London: Collier Macmillan.

5. Ettlie, J. E. 1985. Organizational Adaptations for Radical Process Innovators. Academy of Management meeting, Los Angeles, CA.

6. Connor, M. F. 1980. *Structure Analysis and Design Technique—Introduction*. Waltham, MA. Softech.

7. Mitchell, F. H. 1991. *CIM Systems—An Introduction to Computer Integrated Manufacturing*. London: Prentice Hall International.

8. Chiantella, N. A. 1986. *A CIM Business Strategy for Industrial Leadership*. Dearborn, MI: SME.

9. Majchrzak, A. 1988. *The Human Side of Factory Automation*. London: Jossey-Bass.

10. Pava, C. 1983. *Managing New Office Technology*. London: Free Press.

11. Boulding, K. 1956. *General Systems Theory—The Skeleton of Science*.

12. Porter, M. E. 1980. *Competitive Strategy*. New York: Free Press.

13. Pedler, M., T. Boydell, and J. Burgoyne. 1989. Towards the Learning Company. *Management Education and Development* 20(1):1-8.

14. Wright, R. 1989. *Systems Thinking—A Guide to Managing in a Changing Environment.* Dearborn, MI: Society of Manufacturing Engineers.

15. Browne, J. 1988. Production activity control—a key aspect of production control. *J. Prod. Research* 226(3):415-427.

16. Wolf, W. B. 1964. *Management: Readings Towards a General Theory.* Belmont, CA: Wadsworth Publishing Co.

17. Checkland, P. 1981. *Systems Thinking, Systems Practice.* Chichester, U.K.: Wiley.

18. Harrington, J. 1984. *Understanding the Manufacturing Process—key to successful CAD/CAM implementation.* New York: Dekker.

19. Scheer, A. W. 1988. *CIM—Computer Steered Industry.* London: Springer-Verlag.

20. Chestnut, H. 1967. *Systems Engineering Methods.* New York: Wiley.

21. Wheelwright, S. C. 1967. *Reflecting Corporate Strategy in Manufacturing Decisions.* Business Horizons, February, pp. 57-66.

22. Hayes, R. and S. Wheelwright. 1984. *Restoring our Competitive Edge.* New York: John Wiley & Sons.

23. Porter, M. E. 1985. *Competitive Advantage - Creating and Sustaining Superior Advantage.* New York: Free Press.

24. Skinner, W. 1974. The focused factory. *Harvard Business Review* 52:113-121.

25. Hales, H. L. 1986. *CIMPLAN - The Systematic Plan to Factory Automation.* Arlington, TX: Karen Fine Coburn.

26. Konig, W. 1990. Integration. *Proceedings of CIM-90.* Bordeaux, France.

27. Maniot, J., and G. Waterlow. 1986 A study of state of the art in Computer Aided Production Management. *Proceedings of SERC*, Swindon, U.K.

28. Voss, C. A. 1989. The managerial challenges of integrated management. *International Journal of Operations and Production Management.* 5:33-38.

29. Rockwell, H. 1988. Integrated production. *Professional Engineering.* October:15-20.

30. CEN/CENELEC. 1991. *Framework for Enterprise Modelling.* CEN/CENELEC. Brussels, Belgium.

31. O'Sullivan, D. 1990. Integrated manufacturing systems design. *Proceedings of Factory-2001*. Cambridge, U.K.

32. Ross, D. 1985. Applications and extensions of SADT. *Computer* April:25-34.

33. O'Sullivan, D. 1991. Development of integrated manufacturing systems. *International Journal of Computer Integrated Manufacturing Systems* June 1991:39-53.

34. Jones, A. T., and C. R. McLean. 1987. A proposed hierarchical control model for automated manufacturing systems. *Journal of Manufacturing Systems*. 3:1.

35. CAM-I. 1980 *Specifications for an Advanced Factory Management System*. Gaithersburg, MD. CAM-I International.

36. CAM-I. 1983. *Conceptual Information Model for an Advanced Factory Management System at Work Centered Level*. Gaithersburg, MD. CAM-I International.

37. Cioffi, F., and J. Vlietstra. 1989. Specifications on an open CIM system. *Proceedings of IFIP '89*, APT Nederland B.V.

38. Vlietstra, J. 1989. The need for a CIM architecture. *Proceedings of IFIP '89*, APT Nederland B.V.

39. Bauer, B., J. Browne, J. Duggan, and G. Lyons. *Shop Floor Control Systems*. London: Chapman & Hall.

40. Pingry, J. 1990. HP releases OPENCIM. *CIM Strategies*. VII(7):3-7.

41. IBM. 1987. *Computer Integrated Manufacturing—the IBM Experience*. Dartford, U.K.: Findlay Publications.

42. O'Flaherty, C., and G. Grirola. 1988. *Reference Model for Computer Integrated Manufacturing*. Galway, Ireland. Digital Equipment Corporation.

43. Thacker, R. M. 1989. *A New CIM Model*. Dearborn MI: Society of Manufacturing Engineers, 1989.

44. Ranky, P. 1986. *Computer Integrated Manufacturing: An Introduction with Case Studies*. London: Prentice-Hall International.

45. Krause, F. L., P. Armbrust, and M. Bienert. 1988. Methodbases and product models as bases for integrated design and manufacture, *Robotics and Computer-Integrated Manufacturing*, 4(1/2): 33-40.

46. Rock-Evans, R. 1989. *Analysis Techniques for CASE: A Detailed Evaluation*. Arlington, MA. Cutter Information Corp.

47. Wyatt, T. 1988. Methods and techniques for systems specification. *The FMS Magazine* April:91-95.

48. Mullery, G. P. 1979. CORE: method for controlled requirements expression. Fourth *International Conference on Software Engineering*, New York.

49. Gane, C., and T. Sarson. 1979. *Structured Systems Analysis—Tools and Techniques*. Englewood Cliffs, NJ: Prentice Hall International.

50. Ang, C. L. 1989. Planning and implementing computer integrated manufacturing. *Computers in Industry* 12:131-140.

51. Franks, I. T., and J. Gorman. 1989. A strategical approach to computer integrated manufacture. *Journal of Engineering and Manufacturing*. 203:261-267.

52. Banerjee, S. K., and I. Al-Maliki. 1989. A structured approach to FMS modelling. *International Journal of Computer Integrated Manufacturing* 1(2):77-88.

53. Doumeingts, G. 1987 Use of GRAI method for the design of an advanced manufacturing system. *Sixth International Conference on FMS*, Torino, Italy, 1987.

54. Doumeingts, G., D. Chen, and B. Vallespir. 1989. *Review of Existing CIM Architectures, Methods & Tools*. GRAI Laboratoire. Bordeaux, France.

55. Mesarovic, M. D., D. Macko, and Y. Takahara. 1970. *Theory of Hierarchical, Multilevel Systems*. London: Academic Press.

56. Cutts, G. 1987. *Structured Systems Analysis & Design Technique*. London: Paradigm.

57. Mertins, K., and W. Sussenguth. Integrated information modelling for CIM: an object orientated method for integrated enterprise modelling. *Proceedings TOOLS '89*, CNIT Paris.

58. Vernadat, F., A. Di-Leva, and P. Giolito. 1989. Organisation and information systems design of manufacturing environments: the new $M^*$ approach. *Computer-Integrated Manufacturing Systems* 2(2): 69-81.

59. James Martin Associates. 1988. *Information Strategy Planning*. Dublin: James Martin Associates.

60. Wittry, E. J. 1991. *Functional Analysis - Simplify Before Automating*. New York: Van Nostrand Reinhold.

61. Rui, A., R. H. Weston, J. D. Gascoigne, A. Hodgson, and C. M. Sumpter. 1988. Automating information transfer in manufacturing systems. *Computer-Aided Engineering Journal* June:113-120.

62. EIA. 1987. *Manufacturing Message Specification* Part 1. Washington, DC: National Bureau of Standards.

63. PDES. 1987. *A Report of the PDES Initiation Activities*. Washington, DC: National Bureau of Standards.

64. Brodlie, K. W., and G. Pfaff. 1983. An algorithmic interpretation of the GKS text primitive. *Computer Graph Forum* 2:233-241.

65. Owen, M., and M. S. Bloor. 1987. Neutral Formats for product data exchange: the current situation. *Computer Aided Design* 19:436-443.

66. Kaplan, A. 1964. *The Conduct of Inquiry*. San Francisco: Chandler.

67. Koontz, H., and H. Weihrich. 1990. *Essentials of Management*. London: MacGraw-Hill.

68. Attwood, M., and N. Beer. 1988. Development of a learning organisation. *Proceedings of Lessons from Success* Mead, U.K.

69. Argyris, C., and D. A. Schon 1978. *Organisational Learning: A Theory in Action Perspective*. London: Addison Wesley.

70. Revans, R. W. 1982. *The Origins and Growth of Action Learning*. London: Chartwell-Bratt.

71. Frick, J., and J. O. Riis. 1991. *Organisational Learning as a Means for Achieving Both Integrated and Decentralised Production Systems*. Sweeden: University of Aalborg.

72. Lewin, K., R. Lippit, and R. White. 1939. Patterns of aggressive behavior in experimentally created social environments. *Journal of Social Psychology* 10:271-299.

73. Janis, I. L. 1972. *Victims of Groupthink. A Psychological Study of Foreign Policy Decisions and Fiascos*. Boston: Houghton Mifflin.

74. Grudin, J. 1988. *Why Groupware Applications Fail: Problems in Design and Evaluation*. Microelectronics and Computer Technology Corporation.

75. Hill, P. 1971. *Towards a New Philosophy on Management*. London: Gower.

76. Office of Technological Assessment. 1984. *Computerised Manufacturing Automation*. Washington, DC: Government Printing Office.

77. Nadler, D., and M. Tushman. 1980. In *A Congruence Model for Diagnosing Organizational Behavior*, ed. D. Kolb, I. Rubin, and J. McIntyre. Englewood Cliffs, NJ: Prentice Hall.

78. Pun, L., and G. Doumeingts. 1991. *Integrated Conceptual Reference Modelling in Production Management Systems*. GRAI Laboratoire, University of Bordeaux I, France.

79. Wang, M., and G. W. Smith. 1988. Modelling CIM systems. Part I. Methodologies. *Computer-integrated Manufacturing Systems* 1(1):13-17.

80. Softech. 1981. *Integrated Computer Aided Manufacturing (ICAM) Architecture. Part II: Composite Function Model of "Manufacture Product" (MFG0)*. Waltham, MA. Softech.

81. Harrison, M. 1990. *Advanced Manufacturing Technology Management*. London: Pitman.

82. Riggs, L. R., and G. H. Felix. 1983. *Productivity By Objectives*. London: Prentice-Hall International.

83. Margerison, C., D. McCann, and R. Davies. 1988. The Margerison-McCann team management resource—theory and applications. *International Journal of Management* 7(2):1-32.

84. Schonberger, R. J. 1982. *Japanese Manufacturing Techniques* New York: Free Press.

85. O'Sullivan, D., A. Brennan, and T. Sweeney. 1991. Design and evaluation of assembly lines. *Workshop on CIM Systems Planning Tools*, Saarbrucken, Germany.

86. O'Sullivan, D., P. Fitzpatrick, N. O'Toole, and M. Carroll 1991. Development and operation of integrated flexible assembly system. Proceedings of FAIM '91, Limerick, Ireland.

87. O'Sullivan, D. 1990. Design and operation of mixed and multi-model assembly lines. *Seventh International Conference Irish Manufacturing Council*, Dublin, 1990.

88. O'Sullivan, D., and J. Lane. 1990. Demand pull flexible manufacturing. *Seventh International Conference Irish Manufacturing Council*, Dublin, 1990.

89. O'Sullivan, D., and J. Lane 1990. Development and operation of a demand pull sheet metal flexible manufacturing system. *Journal of Materials Processing Technology* December 24:463-472.

90. O'Sullivan, D. 1991. Development and operation of FMS using structured techniques. *International Journal of Production Planning and Control* 2(2):122-134.

91. O'Sullivan, D. 1991. Project management in manufacturing environments using IDEFo. *Journal of Project Management* 5(1):39-58.

92. O'Sullivan, D. 1990. Integrating CIM subsystems using structured techniques. *Proceedings of CIM-90*, Bordeaux, France, 1990.

93. Browne, J., P. Sackett, and J. Wortmann. 1993. *The System of Manufacturing—A Prospective Study*, Galway, Ireland: University College Galway.

# INDEX